Leading with Imperfect Feet

Leading with Imperfect Feet

How my disability enabled me to become an even more agile leader

< DAVID DAME

< Foreword by ANDREW HILL

an imprint of Microsoft

Distributed and printed by Simon & Schuster

Published in the United States by 8080 Books,
an imprint of Microsoft Corporation

1 Microsoft Way, Redmond, WA 98052

https://aka.ms/8080books

Cover Design by Shyla Lindsey
Typesetting by Nord Compo
In association with Wise Wolf Creative

ISBN 979-8-9937554-0-3 (Paperback – 8080 Books)
ISBN 979-8-9937554-3-4 (Hardcover – 8080 Books)
ISBN 979-8-9937553-7-3 (eBook – 8080 Books)

Table of Contents

Table of Contents

Foreword

When Dave told me he was writing a book, my first reaction was that of gratitude. Gratitude that the perspective and the lessons of his lived experience would be captured in an artifact for many. Now having read the book, I am excited. Excited for you to jump in, zoom out and reflect about leadership and about life through the stories of a man with a truly unique perspective on what it means to live, create, grow, and lead.

I first met Dave when he was new to Microsoft, and we were scheduled to meet for 30 minutes as our teams would be working together. We ended up talking for an hour and a half, and in that time, it became apparent that Dave was much more than someone joining to lead our accessibility design efforts. In fact, that was but a small part of what he had to offer. We were able to connect at many different altitudes and the depth he was able to bring to product making, culture, change management, leadership and innovation were very powerful.

In the time that has passed since then, Dave has become a dear friend, confidant and collaborator helping to craft many of the most important things we work on in Windows & Devices.

I have come to learn that a lot of the depth Dave possesses comes from the fact that he, in a very real sense, lives in the future. He has been using technology to augment his life and his work from a very early age. As we lean into how AI is changing the ways people live and work, Dave is at the vanguard, experimenting and showing the way for how others can take advantage of technologies. I see many people struggle to take advantage of these new tools, but not Dave. This is just one more piece of technology to augment his life and he is a natural. His example scales. I see people struggle with communication and teaming, but not Dave because he learned how to do it from the beginning.

Dave reminds me often that he aspires for those around him not to be inspired by his journey, but only to be inspired by his work.

This book is an example of that work; it is worth reading and being inspired by. Dave has built so much. I am thankful that it is being captured in this book. We should all be excited for the conversations and reflections it will generate.

Andrew Hill,
Microsoft CVP for Surface Product

Introduction

I had no idea what I was getting into when Microsoft hired me as Director of Product Accessibility. I'd been a professional for a quarter century and had lived much longer than that with cerebral palsy, but accessibility wasn't really my thing.

Until I applied to Microsoft, I'd never disclosed to potential employers that I had a disability. Those other firms didn't need to know I moved in a power wheelchair, so long as I wasn't made aware that the interviewer was possibly a close-minded ableist. So, I used my disability as a dipstick to see how progressive a company really was, how willing they would be to make changes to get high performance. (It was quite something to roll into a job interview and have the person resist eye contact and ask questions like "Has your wheelchair ever gotten in the way of your ability to perform professional duties?" It was something else to not respond with "You wouldn't be holding my résumé if it had!")

When I was up for the job at Microsoft, I was also the top candidate for a VP of Agility role at a major bank. I was coming off a five year run at Scotiabank, where I created and implemented performance models for 85 different teams around the world. After going through 12 interviews at the prospective new bank,

I finally received an offer. The compensation figure was at a level I'd typically associate with that of an international drug kingpin.

Conflicted, I asked my wife Kelly what she thought I should do. I could have easily slid into another VP job at another bank. The role at Microsoft, however, which came about after I gave a leadership talk at a company conference, would be a career-180-degree turn.

"You know, you've never taken a job for money," Kelly said.

Then I showed her the bank's offer.

"Wow," she said, her mouth agape.

But her shock eventually gave way to wise-minded reasoning.

"Sure, you can go to this bank and help them make money," she said. "Or you can go to Microsoft and help people with disabilities have the life you've had with hopefully a lot less effort."

Microsoft was offering a product-focused role, as opposed to a full-bore digital portfolio position like the ones I was used to. Microsoft was also based in the United States, which meant I could lose coverage of the home support care I had in Canada. Still, Microsoft wanted me, and Kelly validated the same thoughts I'd had about the possible change.

I felt I needed to be a part of the next wave of technology that was going to help carry more people with disabilities through school and work. Potentially going to Microsoft would be a full-circle moment because that same company's technology helped make it possible for me to get an education and become a professional. Microsoft Word and Excel helped me manage at university, so maybe I'd oversee the development of a program, app, or hardware that would provide the same opportunities for more people like me.

Being geographically far away from my team seemed like a barrier, but then I'd never been afraid of a challenge. It's how I was raised, after all.

Dad was a hard-nosed, working-class guy who owned a small HVAC company. Despite his inherent toughness, he had a way of connecting with people, and he and his team were like a family of their own. (At my father's funeral, one former employee recounted to me that Dad had laid him off, but then brought him groceries until he got a new job.) My father understood that the best leaders deeply understand the people around them, which is how they get their team members to buy into a vision and follow directions.

When it came to parenting me, he preached independence—within reason. He drove me where I needed to be driven, but he also built me a walker out of pipe. On a foundational level, that's not a whole lot different than the experience of an average kid, just with one additional hurdle.

Mom was a firecracker, and despite being about 10 inches shorter and 100 pounds lighter than my father, people were about as intimidated by her as they were by him. My primary caregiver at home, Mom taught me never to give up, to be tenacious and stubborn. This was communicated through her balanced approach to child rearing. Essentially aligned with my father, she provided all the support I required—giving me countless showers, for starters. But she could also be a stern disciplinarian when I needed that in my life. Her message seemed to be: "Your disability will not get in your way of achieving *anything*, but don't think your disability will help you get away with *everything*."

So, my parents didn't necessarily make life easy for me. They knew I was going to grow up and made sure I could fend for myself as much as possible, able to face challenges.

I never wanted to be the token person in a wheelchair at an organization, the one who chirped relentlessly about the needs of *every* consumer across the four corners of the world. I wanted whatever talent I had and skills that I'd built to carry me toward success.

Just as it was difficult to style my hair when my mother got tired of doing it for me, it was hard to oversee teams in Canada, the US, Latin America and India all at the same time. (This was in the 1990s when the internet was in its infancy and video conference platforms were relegated to *Star Trek* movies.) When Dad didn't want to shave me anymore, I struggled with a razor and shaving cream. But I also faced hardship the first time I developed new quality assurance protocol, or an agile system at a company with a multi-billion-dollar bottom line. Actually, it was tough *anytime* I did that because each organization brings its own particular demands. (I guess you can call them "special needs.")

And yet there I was, six months into my employment at one of the most prestigious companies on the planet, filled with uncertainty and doubt, being asked about 3D-printed digital pen grip designs.

Historically, people with disabilities encountered a crossroads early in life: Should they learn to write or learn to type? My personal abilities dictated that I type, while my birth year of 1971 sealed my fate as the only fourth grader in school with a bulky typewriter on their desk.

However, kids with disabilities born in today's world are less likely to make that same choice. Voice-to-type technology has rendered keyboards unnecessary, while other writing-focused innovations, like Microsoft's Surface Pen, paired with a tablet or laptop device, are suitable for a wider range of users. They just have to be able to hold it in the first place.

The pen grips were part of a new line of products designed to enhance device accessibility, and working on them was an opportunity for me to affect real change on the design of a product at Microsoft. Until that point, I'd led a team in Operations, and products were already locked in scope by the time I learned about them. When we moved the team to Design, on the other hand, we were closer to the front of the funnel, and in a product's discovery phase, anything was still possible.

"Well, have you thought about putting a 'T' at the top end of the pen grip?" I said to one of the designers.

"Why?" they asked.

"When it slides out of my hand, the 'T' will keep it from rolling to the floor," I said. "This pen is great until it hits the floor. Then, it's useless until somebody picks it up for me."

Microsoft leveraged my idea and developed a pen grip that looked just like what I'd envisioned. My name was even on the patent.

More importantly, when the grips launched, I felt a jolt of excitement and pride. When a company builds with accessibility in mind, the products end up in the hands of able-bodied people, too. Audiobooks, for example, were initially marketed to blind people, while voice-to-text technology, now a mainstream piece of tech, has been used by disabled people for 30 years. Hearing people watch TV with captions on to better lock into a story, but closed captioning was developed for deaf folks. I'm virtually certain the pen grips have been purchased by able-bodied people who simply appreciate their unique feel.

Let me tell you, the Surface Pen beats a typewriter! Liquid paper was out of the question for me, and in my youth, I threw away countless pieces of paper after making a single mistake on each of them. I realized we had helped someone else have the chance to use a pen for the first time or maybe return to writing after years of reluctant typing. Perhaps they will write a book, too.

This book is the result of ceaseless engagement with heightened personal challenges as a disabled person and tells the story of how this ultimately proved to be a gift that shaped my distinct leadership approach, which I still use today.

Chapter 1.

How to behave like a leader

Look forward to falling down.

I love the beach. The sights, the sounds, the colors. There's something about how the waves keep crashing back to smooth out the sand again and again, after every footstep gives you hope for new beginnings. It's a beautiful place that, to me, inspires creativity and hope.

Unfortunately, sand isn't great for wheelchairs.

We all have barriers as individuals, as groups, as companies. It's not easy, but in some ways, it is easier to just *do* things than it is to *think* about doing them.

Believe me, I know.

When you're born with a disability like cerebral palsy and told all your life what you can't do, you really have to push yourself to see past the limits that others set for you. It can feel like your entire world only extends 10 feet in front of you.

But when you're facing an endless ocean that flows into a ceaseless sky—with no limits in sight—it can give you a different perspective. It's empowering.

So, yeah, I like the beach. But because of the sand-wheelchair dilemma, I generally spend my beach time parked on whatever wheelchair-accessible surface will let me sit safely as close to the sand as possible. That's where I get my view of the water.

Of course, I'm always tempted to get closer, feel the cool waves on my skin. But getting across the sand would be a whole adventure for me, and I usually decide not to risk it.

After all, I might fall. People would look at me. I would feel silly. *Not worth it*, I had always concluded.

But during a Florida vacation with my wife Kelly and another couple a few years ago, something was different. Instead of staying on a platform, we were in the middle of the sand. Kelly and a couple from Sarnia got me into a lawn chair.

The couple told me to give my feet a break. Now, I've always been self-conscious about my feet. I have hammer toes and other deformities. When I took my shoes off, I was so worried about all the imperfections and curves. But when I put them down, the sand wrapped itself around my feet. It felt like the sand was accepting my imperfect feet—and accepting *me*. I was closer than ever to the waves, and I wanted to go into the water.

Most people would get up and walk to the water. Simple, right? For me, simple can be complicated. To be clear, I *can* walk. I even walked a 5K Terry Fox run—perhaps the most physically challenging activity I've ever done. But because of my imperfections, my body doesn't balance without support, and I need help from others. Since the age of 18, I have had the support of health professionals who assist me with basic physical tasks. My wonderful "Team Davey" helps me get out of bed and prepare for my day. As uncomfortable as it can be to have employees—often strangers when they start—perform tasks that most of us consider to be very personal, it was either that or staying in bed for the rest of my life. And I wasn't about to stay in bed.

That's the story of my life. Anytime I want to learn a new function, a new skill, or have a new experience, the cost is that I have to let go of part of my pride and often accept help.

Because I don't do many things perfectly the first time, I must try and fail, often attracting attention while I'm at it.

On that day at the beach—standing on my imperfect feet, which had adapted nicely to the sand—I decided it was time to go into the water.

What was different about that day?

I'd made new friends.

Kelly and I had met the wonderful couple we were travelling with on a cruise the previous year. They didn't know about the decades I'd spent choosing to sit on the beach, but they could sense I wanted to go into the water. So when they urged me to give it a try, I didn't have the heart to say that I'd decided long ago it wasn't worth the risk of looking silly—even though, dammit, I *did* want to go in. I always had.

I need two people to help me walk. One person takes a firm grip of my left arm and the other my right arm. It can look like the people helping me are pulling me, and it definitely doesn't appear as though I'm in control of the situation.

But the couple convinced me. As I got ready to walk to the water, I knew other beachgoers would stare. I made myself smile imagining them wondering why people appeared to be dragging me to the ocean: *Is he alive? Is he dead? Are they trying to get rid of the body?*

It was a ridiculous struggle.

I had never walked barefoot in sand before, and I can still remember the sensation of those first steps. It felt awesome. The warm grains hugged my feet as they sank in with each step. The sand seemed to perfectly form to the shape of my imperfect foot. I felt like it accepted me and supported me.

Still, I wasn't used to walking on sand, and my foot kept slipping as I shifted my weight.

After about 10 steps, I fell.

Flat on my back like a helpless turtle, I looked up at the alarmed faces of Kelly and my friend who were standing over me. "Dave?" they were saying. "You OK?"

Of course I wasn't OK! Everyone was looking at me!

I was mortified. I wanted to go back to my chair and forget this whole thing. What an embarrassment. Maybe I couldn't do this. Maybe I shouldn't have tried.

For a moment, I was flooded with regret. But I'd felt that way before.

After a lifetime of living with a disability, I'd had enough experience with failure to know that an immediate decision was in front of me. I could give up or press forward. If I went back to my chair, I'd be safe from falling again. But I'd also be giving up on what could be—on my dream of walking into the ocean.

The thought of the waves, and the cold refreshing water, and all those years I'd stayed in my chair, wistfully watching others go in, pushed me forward. My reason *why* was greater than the struggle it was going to take.

On top of that, I had confidence in my team.

With their support, I got back up and started walking again. I fell again. And I got back up, and fell again. And again. I was a real life Weeble Wobble.

Finally though, I started to get my groove. I got used to that beautiful warm sand on my feet.

Then, wet sand. That doesn't feel nearly as good. It's clumpy and cold and gritty and sticks to your feet.

But if I wanted to walk into the ocean, I knew I had to go through the unpleasant parts. So I kept moving forward.

Finally: Water! It was a new sensation all over again. When there are waves coming, sometimes you can go with them, and sometimes you have to hold fast and resist. Through each wave, I put all my effort toward taking another step. As I got deeper, it became easier to balance, but at that point it didn't matter. I was completely fixated on my next goal, which was just to get as far into the water as I possibly could.

Knees. Hips. Chest.

"Far enough," Kelly said.

But it wasn't. I kept going.

Shoulders. Chin. The horizon still beckoned.

Had I *really* gone far enough?

I would've kept going, but I knew that if I kept walking, of course, eventually I would drown. So I stopped there for a moment. It was wonderful to see the water close-up, to finally know how it *felt*.

When we turned back, I looked at the shore, where I had spent the rest of my life before that moment. I remembered sitting in my beach chair, looking at the sand and water, thinking how far away it looked. The very idea of walking across that eternal beach into the water seemed wildly audacious.

But as I stared back at the beach from the water, that whole distance looked much less scary. It really wasn't far at all. I could still see my parked chair and towel.

I walked back through the water, up the sand, back to my chair and sat down. The return journey felt shorter. Easier. No big deal at all, really.

Soon, I noticed an older man with a walker making his own journey down to the water. He went in, turned around and came back.

"I did it, too," he said to me as he walked past my chair.

I admit, that felt pretty cool.

I had to ask: "Was seeing me go in the water what made you do that?"

"No," he said wryly. "I just knew that if you could fall down and look like an idiot and get back up, I could too."

I often think about that exchange. It wasn't my success that inspired that man. It was my vulnerability. It was seeing me fall and get back up.

I looked like such an idiot, apparently, that it was awe-inspiring.

This book is about leading others to the water.

Am I trying to say, *You too can inspire others by looking like an idiot?* Not really. But kind of.

The point is that allowing yourself and your team members to be human, trying new things and failing like humans do will set the tone for an atmosphere that will lead to success.

How do I know?

Well, according to the medical experts, I wasn't supposed to learn to walk or talk. I'd never have a so-called normal life.

Guess what? I'm a Senior Director at Microsoft with my own home, a beautiful and supportive wife, good friends and a lifestyle that brings me on vacations all over the world.

I've spent all my years pushing boundaries and changing to adapt to the environments around me that were never meant for someone with my physical disabilities.

In fact, I've turned that into a career helping businesses and organizations find new and better ways to succeed by embracing agility, diversity, humanity, and the risks that come with innovation.

Some disabilities are more obvious than others, but every individual and every organization have challenges on their way to achieving goals. I help people see today's barriers as tomorrow's opportunities.

As a leader, you might not have the skills and behaviors in your toolkit to get you to the ocean today. But that's OK. You can start building them.

Vulnerability is strength

I was 18 years old and heading to an out-of-town university when I hired my first employee.

One of the responsibilities of the job meant my employee would need to help me shower. And that meant, the person who I'd never met before would have to see me naked within days of being hired. Super uncomfortable.

Think of your 18-year-old self agreeing to let a stranger bathe you. That might capture the vulnerability that I had to accept. But I wasn't the only one.

My parents had to allow themselves to be vulnerable, too. After spending the previous 18 years looking after me and managing the logistics of my life—from transportation to personal care—they had to hand over the reins and let me take control to help me grow.

And in my case, that began with trusting me to interview, assess, and make judgement calls about strangers who would soon have my life and my whole naked self in their hands.

Mom and Dad had a lot to lose, taking a risk like that. It must've been terrifying.

But they set an example of true leadership. In turn, I have allowed myself to be vulnerable, to trust my team and to have the grit and tenacity to move forward.

Fast forward two decades and I specialize in helping organizational leaders delegate decision making authority to team leads and employees—to trust their staff to make important calls.

I can tell you that fear of giving up control over decisions is one of the biggest things holding organizations back from change. It makes leaders feel vulnerable, and it makes teams nervous, too, because they often haven't been encouraged or empowered to make tough decisions before.

I learned to be vulnerable and trust others because I had to. I was literally naked. But sometimes vulnerability feels like being naked, and we don't want others to examine our flaws if we can help it. What you find is there's a lot of anxiety. When a new support worker comes in to shower me, I'm in my most intimate sense of vulnerability. I'm on a shower bench, completely at their mercy. If I want to achieve my goal of coming out clean and unhurt, I have to be able to trust they don't use too much pressure and hurt me or pull my skin in a certain way that will cause pain. Imagine explaining to a stranger where to touch you around your personal parts.

Each time I have to remind myself I hired these people because I believed they could do the job. Now I have to trust them to do what they are trained and equipped to do.

If I could learn to be vulnerable and trust my team to help me achieve my goal when I'm at my most raw and vulnerable, then I need to be able to trust my company team to help me achieve corporate goals.

In today's world, leaders need to possess a few important traits: They need to be real, be curious, be tenacious, and be open.

Be Real: Show your imperfections

For years, I didn't use the washroom at work. It wasn't accessible and I didn't want to ask for the washroom to be retrofitted to be made accessible.

More on that later, but the point is this: As leaders, we often try to hide our imperfections.

I mean, you can't hide my wheelchair—the cat's out of the bag on that one. But early on in my career I didn't want to be known as a person with disabilities who had skills. So I was putting a lot of my effort into hiding my disability, but unwittingly hiding my skill at the same time.

I didn't wake up until one of my team members who had caught on said something.

"Dave, it's not that we value you because you don't have limitations, we value you because of how far you've come despite those challenges," she said. "That's what inspires us."

It was an "a-ha moment" for me.

What I thought was going to hinder my brand, drew people to me. I wanted to prove that I wasn't about my disability and so I tried to ignore it or make it invisible, but the fact that I was overcoming challenges every day, several times a day is what inspired the people around me.

Like the man at the beach who followed my lead only after seeing me fall and get back up so many times, my corporate teams over the years knew they could trust me when I said it's ok to stumble and it's ok to fail. Because I put my money where my mouth was.

In the corporate world, nobody wants others to focus on their mistakes or vulnerabilities. But as leaders, we need to remember that success is often the result of a struggle with imperfections. Showing that we are human and make mistakes empowers our team. Your struggles motivate others to put in the work to overcome their own challenges and to move forward one step at a time.

Take a page from Phil Jackson

Take a look at one of the most celebrated leaders in sports: Phil Jackson, who coached Michael Jordan and the Chicago Bulls to six championships.

In media interviews, Jackson has talked about struggling as a player himself. He said he was not as good as his teammates. Yet as a coach, he got the most out of his team, year after year.

Why? Jackson didn't take skill or talent for granted. He could empathize with the athletes when they struggled and knew how to encourage them to do the work to overcome their barriers.

It's about being real, which is about being vulnerable.

Vulnerability is about revealing things about yourself that you have wanted to hide. And once you become open about those things, you realize you never need to be embarrassed again.

People are inspired when they see others struggle and overcome. When leaders show vulnerability by stretching to try something new or pivoting to meet a need without knowing exactly how it will turn out, team members are enabled to take risks.

I think of myself sitting on the beach all those years, longing to go to the water but refusing to make the walk because I'd fall along the way and people would look at me and have *opinions.* If they didn't see me fall, they couldn't call me weak. Right? I think that's the model of leadership a lot of us have in our heads: don't take a risk.

Instill confidence in your leadership by doing things perfectly. Or don't do them at all. Save face. Don't give people anything to talk about.

But true leadership is the kind that inspires others to get up and follow in your imperfect footsteps. It wasn't the sight of me in the water that inspired the older gentleman. It was the sight of me falling over and over again. Those falls encouraged him to give himself permission to try that walk—knowing he might fall too.

To get to the ocean, I had to be comfortable being an outlier—the only one on the beach that day trying to carry out this crazy idea. I experimented as I went through each different stage: the warm sand, the cold sand, and the water. I fell and got back up, slipped and recovered, over and over again. And when I was done, others were starting to catch up; another person with mobility issues felt inspired and empowered to follow my path.

Only, of course, *he* looked like less of an idiot.

Working as a change agent in the corporate world, means I often act in a supporting role like that couple who didn't know I had long ago given up on walking in the ocean in favour of the comfort of sitting on the beach, and who convinced me that with their help, I could do it.

Any company that has gotten comfortable sitting on the beach, so to speak, is disabling itself. Inertia might feel safe, but it takes a terrible toll because it prevents what could be. Sometimes you just need a couple of new people, with their new perspective, to come in and convince you that today you're going to give it a shot. And when you get to that horizon that you never thought you could reach, you'll turn back and see that it was always within your grasp.

Displaying your weaknesses is uncomfortable. It goes against every instinct we have. That's why many people, and many companies, try to conceal their vulnerabilities; it's more comfortable if we don't look at it. But it's a false comfort; it prevents us from achieving growth by tackling our insecurities head-on.

Worse, it can lead to disaster.

Many companies want to focus on growing their strengths. Think of a financial institution with a strong commercial side. That company would probably want to put its resources into developing more commercial products and getting more commercial clients, because that's what they're good at.

But building on your strengths is just a defence mechanism. It preserves what you already have; it will never give you something you've never had. And it doesn't inspire anyone.

What's inspiring for people is to see someone stretching themselves, trying something new, pivoting to meet a need but without knowing exactly where they'll end up.

As you empower and accept the individuals within the company *as individual people*, the entire company becomes more human and more responsive. To really be innovative, you have to have a culture that allows for small, affordable failures. This should be easy, because when we're dealing with individuals, we know no one's perfect and people are going to make mistakes.

If you are a people leader, your employees probably weren't around when you were starting out. They didn't see what you went through to rise to your current rank: the mistakes, the pain of losing face, the public recovery. It can be easy to mask, now that you're a leader. But if you do, you're missing an opportunity to connect with your team. To be real and allow them to be real. Worse, if they think you don't embrace imperfection, they may sweep hide theirs from you, sweep mistakes under the carpet and avoid taking risks. And as shown above, that's how disasters happen.

Be Curious: Wonder. Ask. Learn what makes others tick

Back when I was a kid, before I started inspiring people, I got bullied sometimes. I mean, it was the 80s, who didn't, right? I really noticed it when I started going to a "regular" school in Grade 4, once my parents convinced educators that my wheelchair didn't affect my IQ and I could more than keep up with my peers in a regular class. I remember being so excited to go there, thinking the school playground would be a great experience.

Growing up in a very rural area, there were only two other kids around my age. My neighbors Danny and Dave always played with me and it was a blast. But the kids I went to school with had lots of options other than me. They didn't know how to play with a kid with cerebral palsy.

And that's when I learned what was worse than being teased. Being ignored. Your first notion when you're left out like that is that you're a loser. But I was curious. I started to become interested in the other kids and I realized that by watching them, I was learning about them and learning how to connect with them.

In a way, it started because I needed to entertain myself. But observing kids on the playground taught me so much that I've taken with me throughout life. Games, schoolyard fights, disagreements, laughter: they're all driven by interpersonal dynamics.

I was starting to see how communication and relationships happened. I learned about leadership and collaboration and conflict and negotiation.

And then I started to use what I learned as a way to connect with the other kids.

I'd hear someone talking about stickers. And later I'd be like, "I like stickers."

And sometimes it was as simple as that.

As I got older and started to form networks in other settings, my curiosity and observation continued to spark connections and build bridges with peers, colleagues and employees.

I have always had to change and pivot at a different cadence from others. It puts me in perpetual "discovery mode," in a constant space of innovation.

In order to participate in a world that wasn't built for me, I'm constantly asking questions, exploring, learning and trying new things.

Sure, I have to do this. But it has served me well and led to leadership opportunities that would have been beyond the wildest dreams of the adults who would've seen me in the schoolyard at 10 years old, watching the other children play.

Curiosity is essential to relationships.

As leaders there is nothing more valuable than the relationships we maintain with the people in our personal and professional lives. Stay curious about your customers and your employers, and you will.

Be Tenacious: Fail *and* try again, fall *and* get back up

Business is about people, and the thing about people is they fail. They disappoint, misunderstand, underperform and sometimes don't even show up.

As a result, business organizations should anticipate mistakes and problems, and be ready to face challenges.

The key is to overcome hurdles and recover from failures quickly, then apply the learning for next time so that whatever your organization is putting out doesn't get completely messed up by the challenges within.

Think about what a thermostat does. As the temperature in your home gets too hot or too cold, the thermostat inspects, adapts, and corrects. You experience a stable temperature, but the reality is that constant adjustments are being made.

In an organization, it's not so smooth. Those constant adjustments require tenacity—the grit and determination to keep striving for the desired output.

In my work, I help companies become more agile, but I also tell leaders to use that framework when they are mentally approaching situations.

In an agile approach, when companies encounter a failure, they don't skip a beat. They collect feedback, identify what was learned and build the next version of the product or redesign the plan and then try again.

It becomes a cycle you can identify. When you know you're going to try and fail in cycles like this, you're ready for it. By the time you see the current version is starting to fall short, you're already thinking about how you'll build the next one.

That's tenacity.

That day on the beach, I would have *liked* to make it all the way into the water without falling. But I failed and that was ok. I made adjustments to the way I was standing to balance myself and feel again. Then I made more adjustments and the team supporting me adjusted the way they were propping me up and then I fell again. I learned how to walk in dry sand and wet sand, and on my way back, I fell less. I didn't gain progress by sitting in my chair running analyses on how to walk in mud.

I did it a more efficient way: through trial and error.

Tenacity will help you achieve goals in an uncertain or challenging environment. Tenacity is about trying and trying again like I did on the beach. And as a leader it will give you the drive you need to keep encouraging your team to do the same.

Tenacity is closely linked to. . .

Grit, which can take you from feeling weak and ashamed to feeling brave and vulnerable. It's about persevering through an uncomfortable process; being comfortable with rough patches in the road. A leader with grit is willing to push ahead with a concept that is incomplete but has potential. It may not be pretty just yet; it may not be perfectly smooth. But you stick with it and make improvements and learn. Think of a piece of sandpaper.

Through the grit of the paper, you can *feel* where the imperfections are; you can find them and work them down.

Most often, grit is about forcing people to slow down and recognize there's a problem. I see this often; organizations enter a growth phase and just want to scale up in the blink of an eye. They create this Wild West where the company doesn't have the culture or the policy framework to handle their new capabilities, but they don't want to hold back because there's a business opportunity. Anyone who tries to raise an issue is told they're making a "career-limiting move".

A person with grit knows the problem isn't going to go away. Sure, you can sand around it and keep your speed high, but you'll have to address it sooner or later. A company with grit would rather focus on that problem and make sure it's solved as soon as it comes up, than gloss over it and pretend it's not there. Because the problem isn't the enemy; it's just a rough patch that hasn't been worked through yet.

There are learnings there; there are knots and bends in the wood that will lend character to the final piece. But first—you've got to work it out.

I have to admit that sometimes I need a bit of grit just to get me up and ready for work in the morning.

For my job, I need to dress for a business environment, including shoes, socks, a dress shirt and pants. That's my goal; a business-appropriate outfit.

To get dressed exactly the same way that others do, then I'd have to do it all by myself, in about five minutes (with no swearing?).

But I can't do that. So in order to reach that goal that I share with thousands of other people, I have to figure it out for myself. My way involves a personal support worker, a significantly longer period of time, and only some swearing.

By the end of the getting-ready-for-the-day process, I'm showered, shaved, teeth brushed, and dressed like most people, but to get to that status, I have to navigate different challenges with help from staff.

I get through my day by constantly adjusting and pivoting to meet my challenges. The end goal is the same as that of my colleagues: put in a good day of work.

I just have to discover my own "best practice" of getting there.

And then be tenacious in my approach. Because if you are tenacious, you will not lose sight of your final destination.

Be Open to change

Open-mindedness is the ability to call yourself out when needed. It's a mindset where all of your assumptions are on the table; you're continually testing your own limits.

When I stood up and started my walk down the beach, I didn't really know if I could do it. But I was determined to find out.

Open-mindedness is also about creating an environment where it's safe to ask for help. You have to model that as a leader, by letting your employees, peers and supervisors see you admitting that you need help with something. Asking for help can feel so painful, like admitting defeat. But when you ask for help, you're revealing that there's a problem you can't fix. That's valuable knowledge, and it's the first step to finding a creative solution that benefits others and solves other problems, too.

Asking for help is also an important bonding tool. We often think that the best way to bond with others is to do them a favor. Now they're in our debt. That's the best place for people to be, right?

But in fact, studies show that people feel more bonded to you when *they* do something for *you*. By that logic, the best way to make a new friend is to ask that person for help.

When I talked about vulnerability, I said that it's about recognizing the power of your so-called weaknesses. I mentioned how Boeing and Facebook shied away from admitting their weak spots and are still losing the trust of their customers.

Exercises

1. Make a list of the things you try to mask or hide when you're at work. Behaviors, insecurities, weaknesses. When you're presenting in a meeting and you're hoping you don't "slip up," what does that slip-up look?

What's the one thing you're hoping you don't do? **Make a list.**

2. Now, think about coping strategies you've developed to help you hide these things. What behaviors have you built to compensate for your weaknesses so no one sees them? **Make a list.**
3. Now, imagine for a moment that some magical event has taken place and your company has wholeheartedly embraced all of your weaknesses. They know everything and they love it. Really take a moment to take in this new freedom you have. Now, what kind of progress do you think you could make at work?
4. Now, think about your team. If you had to guess, what are some of the disabilities your team members are probably trying to hide from you?
5. What questions do you think you could ask to invite them to share these vulnerabilities with you?

Chapter 2.

Change from the bottom up

Build a diverse team and reward feedback and new ideas

One thing you should know about me is that a big part of every milestone in my life has revolved around a bathroom strategy.

When I was a kid, every new place came with the question, 'How is Dave going to go pee here?' It was like my achilles heel. Back then, the adults would find a volunteer to go with me to the bathroom, to make sure I did my pants up OK. My mom even sewed these pants with a Velcro fly. Her heart was in the right place, but it got awkward when puberty hit.

But how Dave is going to pee has been the thread throughout my journey. Anytime I started anything new, from a new elementary school, to university, to a job, I had to make a plan about how and when—and sometimes *if*—I would relieve myself. My first job after college was no different.

Unless you define "different" as worse. Then, well . . . yeah, it was different.

To set the scene, the job was in an ad agency in the 1990s. It was definitely an alpha male environment. For instance, shortly after I started, some of my colleagues nicknamed me "Speedy," because of my wheelchair. All in good fun, I'm sure, but as the guy in the wheelchair, it didn't feel very fun.

However, my job hunt had taken so long. By then, I'd probably sent out 200 resumes and only gotten a few interviews. In some cases, after securing an interview, I wouldn't be able to access the building or the office. I was determined to make this work. So determined that I had already resolved to work four extra hours a day because it took me that long to accomplish the same physical data entry task as my colleagues. So, when I found on the first day of the job that the washroom wasn't big enough to fit my wheelchair, I decided not to say anything and just avoid going to the washroom.

For. Three. Years.

Normally, I go to the washroom every couple of hours, so to get through 12 hours a day without a washroom break all of the sudden, I had to completely change my habits. Every week, on Sunday afternoon, I would start limiting my fluid intake just to be safe. Then on Monday morning and every other weekday morning, I would use the washroom at home at the last possible second before leaving for work. During the day, I would carefully manage my fluid intake and would try not to get thirsty by avoiding salty foods and unnecessary talking.

(Little known fun fact: Talking can dry out your mouth and make you feel thirsty)

The upside (of course there's an upside, this is a motivational book on how to make change work for you!) was that I became an expert in **planning and strategy**. From then on, there would be no detail or potential obstacle too small to consider.

My work struggles began to turn around once I got a computer.

Specifically, once I got access to the new-at-the-time Internet. That was a game-changer for me. I could see early on how the computer could help me—and the company—achieve specific goals.

I was the first employee to learn to incorporate it into my workflow. I could have the computer figure out ways to use its processing power to reinvent the ways I was troubleshooting tasks I could not physically do.

At the time, I was running logistics for training courses across Canada. I had to set up venues, catering, and course schedules for people who were registering for activities, events and courses. Before the computer, I would manually schedule and arrange the information, which took a very long time for someone who isn't able to write or type at average speeds. There had to be a more efficient way. Other employees weren't as motivated to find it, because manual inputting hadn't been a challenge and seemed to be working well enough. They could do the necessary day's work in eight hours. It took me 12, but that didn't affect them.

So on the weekends, I started creating a self-serve website to allow people to enroll and sign up for courses, choose their meal preference and location they want to go to, they could do it automatically on their own. Once the online data program was complete, it took me three hours to do the job my peers were doing in eight.

When others saw that the guy who used to take 12 hours to get through his work was getting it done in three, they took notice. The supervisors wanted my advice. Suddenly, my logistics job transitioned into corporate training; people needed my skills and my eye to reorganize their work for computers and the Internet.

This was my first experience demonstrating to a boss the benefits of having a staff with cognitive diversity and life experiences. I encountered a problem before my peers, so I solved it first. My solution became a forward-thinking plan which was rolled out to the rest of the company to make workflow more efficient.

If that alpha-male, relatively homogenous agency wouldn't have hired someone with that disability they wouldn't have got that efficiency they rolled out to everybody. That was the start of me being a so-called "change agent"—of really being able to see the world differently and experiencing an environment mismatch before my peers

Stick with me, this is all leading back to the bathroom. In a good way.

After I showed the company how teaching people to use technology could help achieve their organizational goals in a more efficient manner, it became my schtick. The company moved me from department to department, and in each new section I was tasked with helping leaders find new or better ways to use technology.

It wasn't only a promotion. They also offered me any support I needed. Did I need a team? Another computer? Technology? I had learned what my value add—my secret sauce—was. And only then, did I start to feel I belonged. I was valuable and therefore I was safe.

So when my boss came to me one day to congratulate me on my success and asked if there was anything I wanted/needed from the company, he was expecting me to ask for more staff or more computers. No way. I finally felt safe to speak up. I felt safe enough to say I would like an accessible bathroom.

If you've ever felt like you stuck out in a workplace because you were different, you may know the feeling. I knew people around me doubted my value; they assumed I was some sort of diversity "token" hire. When you're the employee with the difference, you can't shake that feeling. The last thing you want is to draw *more* attention by asking for accommodation. All any employee really wants is to feel that they're safe and valued and I didn't want to be the squeaky wheel. It was too much of a risk

And this wasn't now. It was the 1990s.

But once it was established that I had a unique skill set, and I felt valued, it made more sense to ask for accommodation. I needed a washroom to be able to focus fully at work. And now, I felt confident to ask for one, because I wasn't asking the company to spend $20K to accommodate someone struggling to do the job.

My boss's eyes grew wide with shock, when I told him I hadn't used the bathroom in three years. He asked why I never brought it up before. I told him the truth: "It took me so long to find my first job, and I was struggling to do the job as it was, so I didn't want to create yet another inconvenience."

But now we both knew I was a valued employee and it was in the company's best interest to keep me. They were converting the bathroom because they needed me.

It takes a diverse ecosystem to raise a change

This chapter is about bottom-up change, but before you can look to your team to lead, you need to ensure your team brings a diverse range of perspectives.

In other words, you need to do more than tolerate diversity on your team, or even embrace it. You need to **actively seek out diversity**.

As a leader, your role is to create an employee ecosystem so your team can function at its best. That means looking at how each member fits into the whole rather than treating everyone as equal, individual contributors.

In the above example from my first job, my manager could've said: 'Great. That works for David and helps him do his job like the others.' Instead, he looked at how the system I'd created would benefit his entire employee base. And he recognized how my unique way of thinking would benefit the team and the organization as a whole.

And if he hadn't hired me—a so-called "diverse" employee—in the first place, his employees who were dutifully working to keep the status quo would not have needed to find the efficiencies that benefited the entire company.

But to see those benefits, leaders need to untangle themselves from the traditional, top-down approach to management, through which leaders tell employees what to do and that's that.

After that conversation, my manager did something that should be an example to managers everywhere. He listened to me and used what he learned from my unique line of thinking to lead the company in a new direction.

That's leading from the bottom up and its success required two elements:

1. A diverse team (My disability gave me a unique perspective on problems and solutions)
2. A manager who was willing to listen to someone on the front lines (me) as the expert.

Essentially, as a manager, you should expect everyone on your team to be smarter than you in their own area. Then accept their expertise. You should look to them and listen to each of them to help you make decisions. As a leader, you keep them aligned toward the overall goal, and give them a space to experiment and learn.

To be blunt, an office full of able-bodied white males is not an ecosystem; it's a monoculture.

That doesn't mean hiring anyone who looks different from that model, just so you can say you hired someone different. It means hiring qualified, skilled people who don't meet the model and will add value.

Hiring with a lens toward diversity is about setting up the conditions so that every single person you hire, no matter their background or abilities, will 1) get a seat at the table and 2) feel safe and empowered to speak up and keep their diverse perspectives.

The powerful Black Lives Matter movement which swept North America after a white police officer in Minneapolis killed a Black man named George Floyd, led to an institutional and corporate reckoning in 2020. In the months that followed, many Black, Indigenous and other People of Color shared stories of systemic racism in the workplace.

Their voices and their stories opened eyes across the white-able bodied, heteronormative corporate world of North America, and also opened the door for others: Members of the LGBTQ community shared their stories of discrimination. So did people with disabilities.

What came to light, was that lot of that discrimination has been happening in a covert- under the surface way that made "diverse" employees feel like token hires who were not invited to bring their diverse perspectives to the decision-making tables. Software programmers with physical disabilities overlooked for promotion because supervisors want to avoid "accommodations," Muslim reporters barred from covering events involving a mosque because they would be biased, Indigenous scholars overlooked for promotion, because their expertise and practices didn't align with colonial standards.

These are not only human rights violations, they hurt corporations.

So let's be clear. When I talk about diversity hiring today, what I'm talking about is broader than covering off the legal grounds of discrimination.

To thrive, your organization needs diversity of lived experience. You will *benefit* from including perspectives and ideas, not only of people from different backgrounds, races, ethnicities, genders and sexual orientation, but of people who struggled in school; people who have traveled all over the world; people who struggle to fit into conventional workplaces like I did. You need people who have had life experiences that help them anticipate problems, that make them self-aware, curious and resilient.

You need people who see the world differently.

Can we say something here about how people with disabilities were among the most prepared and have been highly productive team leaders since COVID-19 required so many to work from home? This chapter is definitely a good spot to talk about that stuff, because of the ecosystem focus.

In the business world, many players come from a place of huge privilege, where going through school and getting hired is a rite of passage; some never doubted there was a seat at the table waiting for them.

But for others—particularly members of diverse communities—it is a hundred daily fights just to get to that table. And there is a lot of emotional maturity, insight and experience gained in those fights, and as a leader, you want that valuable perspective on your team. It gives you an edge on the competition. I know.

When you add these outliers to your team ecosystem, you'll get a team with grit, able to get a project over the finish line no matter what kind of unconventional thinking it requires how/why this is an assumption?

Not just that kind of diversity—professional diversity too

I was nervous during my first few months at Microsoft, only partly because my role there was a new one for me. It's also a big deal working at a company with its level of global influence and brand recognition. I wasn't immune to some intimidation.

But over time, it became clear that at least one of my natural skills was a key to success in the position.

I've always been a natural storyteller, which proved helpful during product launches at Microsoft. I won over important people targeted by our marketing department—influencers, journalists, critics, etc.—with stories about my experiences living with cerebral palsy and engagement with technology. Taking this approach not only improved the devices we issued, it injected humanity into the mission to sell products. (I often tell executives at Microsoft, "I have cerebral palsy, but my money doesn't," revealing that disabled people are a worthwhile demographic to tap into.)

The Design team were close to delivering the Surface Adaptive Kit to market. Basically, it was a collection of stickers designed for blind people to locate specific keys on their device. It also helped people with physical disabilities to open laptops and carry laptops. But describing the ways the kit could help someone like me, with stories about the difficulties I faced using laptops, opened up new marketing possibilities for it. Those discussions also led to the addition of a lanyard on the backs of Surface Pros to make it easier for disabled people—and, really, anyone—to unfold the stand.

To find the right mix of employees, start by mapping out the skills and background you have on your team. That's your "matrix of the status quo." Then ask what other skills and backgrounds could contribute to your problem. What don't you have? When you create a global product or service, the only way to ensure it meets the needs of diverse customers is to have diverse people creating them.

It's already happening.

In recent years, tech companies have looked out of the box, they've hired journalists to help them tell their stories, which is helping them become more successful. Banks have hired behavioral scientists to help them understand rationale behind people's financial decisions.

If your team is mostly engineers, what would happen if you added in someone with a fine arts background—maybe an early adopter or particularly tech savvy artist? If you're a bank who hires primarily from other banks, what would happen if you started including people from healthcare or software? What new ideas would you get bouncing around?

Your team may have imperfect feet and that's a good thing

The teams I worked with in that first job after college were made up of typical humans.

They hadn't integrated their day to day to the new computers and Internet phenomenon as fast as I had, maybe because they didn't need to, or the problem was not as personally challenging for them to look for a new way of using technology. But that didn't mean they were stuck.

When I used the internet to change the way I worked, it gave me fantastic efficiencies compared to them. Suddenly they were the ones falling behind in their work by comparison. Me having a new ability to do work faster that they couldn't do without knowing the technology gave them a disability.

Dealing with disability came to me naturally, so that's why the company brought me in to help others integrate technology in their role of day-to-day activities.

The shoe was on the other "imperfect" foot, and because of that, the team grew stronger.

Back in the 90s, nobody was talking about "Organizational Change Management."

But that's what I was implementing, and at the same time I carved out a role for myself that hadn't existed in many companies. As technology quickened the change of pace, and companies needed to be nimbler, that role I created became more important, and more common.

By the time the term "change agent" was becoming widely used in the early 2000s to describe someone who leads change and challenges standard conventions within an organization. It builds the trust. I already had several years of experience—which was not easy, nor sometimes pleasant to obtain.

Coach people on how to build and maintain relationships.

Another skill of mine that has been useful in leadership is that I am eager to engage with people. In other words, I'm a talker. But, more importantly, I'm also a listener, which makes me a more effective talker.

Many people enter conversations with disabled people, I think, a little guarded. They might not want to say the wrong thing and perhaps anticipate disjointed conversation due to a physical impediment. I have mild trouble with speech, but I've never let that take away from my affability. I tend to be upbeat and funny, which really disarms people and makes it much easier to get to know them and build a relationship. Remembering the way my dad ran his team, I know that strong, intimate connections with colleagues are crucial to success.

At Microsoft, I have had the pleasure of working alongside a person with extensive experience in accessibility product design. They're not disabled, and yet they probably know more about disability issues than I do, even those associated with cerebral palsy.

They've helped bring to market some incredible innovations, and yet, when I was hired by the company, a number of this person's co-workers told me they didn't like them. They complained this individual was impatient, gruff, and abrupt.

I had a number of meetings with the team member and it was abundantly clear that this person cared deeply about accessibility. Their heart was certainly in the right place, but they struggled to communicate effectively with people who didn't share their same enthusiasm or background knowledge. Their expectations of others were simply too high.

After piecing all this together, I began coaching this colleague on building better relationships. I told them that was a more effective way of influencing people.

"Do you want to be the smartest person in the room or the most impactful?" I asked.

They got it.

We sat in meetings together and afterward I'd ask this team member how it went. Based on their answers I gave them tips on how they could communicate more effectively. "Listen with curiosity and ask questions" was one key piece of advice I provided.

I also suggested that they perceive themself as a teacher, to educate their colleagues on the necessity for accessible design, on a case-by-case basis if need be.

This person is not only still at Microsoft, four years later, they've been promoted multiple times. They've produced instructional videos and developed a more empathetic ear.

In an age where virtual or hybrid work arrangements are becoming normalized, cultivating strong relationships, while as important as ever, is more challenging. If you work onsite, it's easy to bump into someone in the hallway and ask if they want to grab coffee or lunch. Without any chance of that happening if you work from home, go out of your way to schedule powwows over a video or phone call. You don't want to overdo it—because people can get burnt out on talking with others through such means—but make sure some of the conversations veer toward appropriate and comfortable discussions about personal subject matters, no different than it would be in the office.

The first few months of my time at Microsoft I did this to an exhausting length, but it was worth it. I got to understand what motivated my new team members and built personal connections, all through remote connectivity.

With that said, if possible, you should make an effort to connect with colleagues in person, or have your remote workers come to the office from time to time.

One day, still during COVID times, early in my employment at Microsoft, my superior asked if I might be able to fly from my home in Eastern Canada to Redmond, Washington. They wanted me to co-star in a corporate commercial for the Surface Adaptive Kit. It was important to have my presence, my voice, my stories in the ad to illustrate the helpful capabilities of the product. Despite all the challenges that would inevitably come with the trip—having to get tested for the virus, bringing a helper with me, flying across a continent, with a layover, etc.—I agreed to go.

It wound up being the trip from hell, with a missed connection after a delay, which required a hotel stay and triggered a need for an additional COVID test. My final flight was hit with another long delay, unfortunately because someone on the plane was struck with a serious illness and had to be rushed off the aircraft into an ambulance.

By the time I was at the video shoot, I'd had about eight hours of sleep over the previous three days. And like most video shoots it was expected to last many hours, with various scene takes, etc.

I pulled from every energy source I had to deliver what they needed, and was ultimately successful in my performance. I only got to spend a few hours with my new team, but they were enough to build significantly deeper connections with them than I could have remotely. It all somehow started to feel "real" at Microsoft for the first time for me, and my colleagues got to see firsthand how impactful accessible-forward products can be on the life of someone like me.

How to build diversity. Turnover isn't always a bad word.

Does every person on your team need 10 years' experience?

When a job candidate has a history of changing positions or even industries every two or three years, some managers might see commitment issues.

I see someone who wants to keep growing. They stay in roles long enough to get through the steepest part of the learning curve, and then they want to move on and learn something else. Employees with this type of history crave growth and learning. They can refresh your whole team's skill set and perspective if you let them. Yes, they might leave—but when they do, they'll create an opening for you to bring in new perspectives again.

The traditional mindset for hiring is to recruit employees who will stay the longest. Of course, there is real value in long-tenured employees with institutional knowledge. In fact, I would argue that sometimes retraining a strong employee to learn a new skill set is more effective and efficient than seeking out a new employee who is already equipped with the experience you need.

But that doesn't mean your whole team needs to be at that level or on that track. Consider treating your team like a landscape: you need a couple of trees, with deep roots and leafy branches, to hold the soil together and shelter other plants. But you also need some perennials and annuals; some plants that are there to stay, and others that are just there for a couple of years. Each employee and length of tenure, has something to contribute to the team. Together, they create a beautiful landscape that has something to offer for any season or weather.

Bottom-up, Talk-down, Meet-in-the-middle

Where does change come from in your organization?

When I ask that question during seminars, most leaders say something like, "Executives make strategic decisions, and then senior leaders filter that down in terms of what our teams need to deliver. Managers make sure the work gets done."

But I didn't ask how the company achieves change. I asked *where it comes from*.

In every company, the change really comes from the bottom. That's where the front-line employees are, who need room and autonomy to change the way they work in order for the organization to achieve the larger strategic change.

You know what else is at the bottom? Customer knowledge. Teams with individual contributors and customer-facing roles are the ones who spend hours every day sitting in front of your customers and users, finding out what they want and need.

When I start a new job, I often start by getting to know the front-line workers. They are the ones who are happiest to accept my meeting invites, after all.

Then, I bring their insights up to management.

The idea is to get the two sides to meet in the middle.

Because while fear and anxiety travel from the top of the organization down, innovation travels from the bottom up.

What does it mean to say that fear and anxiety travel from the top of the organization down?

If you've ever sat in a board meeting while an executive excitedly launches some relaunch or new initiative based on the latest motivational book, you might disagree that change comes from the top. Clearly, these Execs are the ones calling the shots… because there's no way you'd be implementing this change if *you* had any say!

That's the old model, and it doesn't work anymore.

Today, the smartest and most successful companies put their people first.

The middle and top must learn to respond to what happens at the bottom

Putting people first doesn't just mean employees; it also means customers. In fact, for a modern company, those lines should start to blur.

As I mentioned, innovation comes from the front-line employees. That's where all of the user feedback is collected—where the client experience is witnessed firsthand.

However, senior leaders are also tracking market, competitor and buyer trends that the front-line employees don't have access to.

It's important to combine all of this knowledge (from leaders and front-line workers) because that's how a company will get a fulsome picture.

A corporate buyer (who is on the leader's radar) might purchase software that a front-line IT professional (who communicates with the front-line worker) will end up using. If the software underperforms, the front-line customer service employees hear it first and should be feeding that information up the ranks, so high level managers can strategize about how to navigate the issue.

When strategy is being decided upon, senior management should provide the "North Star"—the general vision and direction for the company. But an effective strategy needs input from the bottom. In an ideal world, those inputs would collide, and a collaborative solution would be reached, with long-term direction from the top and "boots on the ground" intel from the bottom.

If an organization is constantly learning internally by listening to its employees, and externally by listening to its customers, you'll be able to see market shifts coming and swiftly adapt to meet them.

The final part of the change equation is that in your employee ecosystem, change must be an experiment, where failure *is* an option.

Put safety precautions in place so your team can't blow up the business, and then let them loose.

Respond to customers rather than dictate to them what they want

Every company exists to serve consumers. That's the conventional wisdom. But in fact, we know that most purchasing decisions are made emotionally. That means we're actually in the business of serving consumer *emotions.* Every business today is in the business of building client relationships.

Most companies today are serving a global and diverse marketplace. That means, just to get started on building a relationship, your workforce should reflect your customer base.

In a "bottom-up" company structure, where employees and customers are being listened to, the best way to get insider client information is to hire people who are also your current customer or who represent the type of customer you want to attract.

One especially important kind of diversity in this respect is age. Many companies target specific age groups for their clientele. But we are all getting older! Your current client base is aging, and new customers are entering the market all the time. To stay abreast of all of these different demographics and what they mean to your business, the best place to look is within.

Your long-tenured employees know your current clients the best. They've also been through more cultural transitions; they know what it takes to get a whole company to switch from manual forms to computers to the cloud. Meanwhile, your new employees can often come to represent the clients you want to attract in the future. They understand the coming market in ways that more established employees may not.

As a company, you have this treasure trove of information available to you through your own diverse workforce. And the more diverse your workforce it is, the more representative your data is.

Once you become a leader in a corporation, it's easy to become focused on the solution the company is offering to its clients, rather than the original problem the company was founded to solve.

That's unwise because every solution will become obsolete, but problems are forever. Remember VCRs? They were a solution to the problem of home entertainment. They're obsolete now, but the home entertainment industry is stronger than ever.

The company and the problem exist in a vicious cycle. We need a marketable solution so that we can fund the company. But we need the company and its employees to come up with a solution that's good enough to go to market. Many companies see this as a linear narrative; long ago, there was a problem and they solved it. Now they're resting on their laurels, rolling out basically the same solution as the same company every year.

But I see it as a cycle. There was a problem, and your company formed to solve it. The problem transformed, so you transformed too, and offered a new solution. The problem will change again, and so will you.

In fact, your company is never not changing. That's why it's called *turnover* when employees come and go; like a landscape that's being planted, the old soil is being turned over to get ready for something new.

In fact, your workforce is one of the biggest organic sources of change in the company. To implement transformational change, you have to harness it. That might mean hiring younger, more diverse employees to better reflect your changing customer base. It definitely means listening to your entire workforce and collecting information on how your product is being used.

That's why your employees should feel empowered to "bring their whole selves to work." It's not just another engagement slogan. It's actually a pragmatic statement about user data. Your employees have vast amounts of information about what it's *really* like to use your product and interact with your company, from their unique perspective. You need them to bring that information to work and share it with you. But that doesn't happen without trust, and employees feel trust when they see that they're safe and valued.

But you can also dictate to customers what they want—when they don't know something they could use exists

Early in my time at Microsoft I had deep discussions with Marketing about accessibility products that were in various building phases, and I realized they were taking a "Field of Dreams" approach. They believed "If we build it, they will come." They also seemed to think that disabled people had someone from the medical community who kept up on such developments by our side 24/7.

To their credit, though, they were only just becoming familiar with their target audience for such items, learning about important subjects like Inclusive Marketing.

"As a disabled person," I told them one day, "I honestly don't know how to find stuff I need. I usually happen upon products, seeing other disabled people using them and thinking, 'Oh, I could use that.'"

So we needed to make these accessibility products discoverable, and it seemed to me the solution was to market them with story-driven campaigns. We were going to show disabled people utilizing the items with a tone that communicated "These are products disabled people will *choose*, not just simply *use*." With that, Microsoft wouldn't come off like a company saying disabled people should be grateful to be served by us. Instead, the company would communicate that we're happy to have their business.

So many ads for accessible products were basically inspiration porn up to that point. Even Microsoft had a Super Bowl ad in 2019 for its Xbox Adaptive Controller where disabled young people play video games presumably against off-camera able-bodied kids. At one point a teary-eyed dad said of his disabled son, "He's not different when he plays."

This messaging worked well at the time for our gaming audience, but I wanted to help push the company forward in its messaging, so that we didn't talk down to our customers. We needed to take a different approach with productivity products focused on an adult disabled community.

"Let's get real crips doing real shit," I told a Marketing Director one day—referring to "crippled" folk, not the violent street gang. The mindset should be to treat members of this demographic like we would any other user.

My half-rhyme worked. When it came time to launch new generative AI tools, the Surface Adaptive Kit—the aforementioned sticker collection for devices—and a new line of Adaptive Accessories—including the 3D-printed pen grips—the ads felt much different. While they highlighted that these product users had additional needs, the theme was more focused on, as I said in one video, "empowering everyone on the planet to achieve more."

Exercises

1. Take inventory of your current team; create a matrix and populate it. What are members' levels and areas of experience? What are their professional backgrounds?
    a. Set aside some time to really sit and consider your matrix. What skill gaps are there? What personality traits are you missing?
        i. To get some more insight, go through your team's latest performance reviews. What growth were your employees looking for? What skills did they

want to build? Chances are, your employees have already identified a gap or two.

ii. As you look to fill that gap with a future hire, consider that not only can your new hire bring that missing piece to help your team with their deliverables, but they will also be able to mentor the others in this desirable skill set.

b. Commit to being pushy about getting more diverse hires when you're working with a recruiter. Recruiters will bring you candidates from the same pool (looking at you LinkedIn), with similar backgrounds and career paths. If you want better, you have to push back. Set the expectation that you want something different—a different education, a lateral move, an Arts background instead of finance or engineering—and you won't hire until you get it. If all else fails, use a website like Meetup.com to go out on your own and find the unique talent you want.

2. Imagine someone who is the polar opposite of most of your employees. They're all engineers? This person has an arts background. They're all aspiring managers? This person is an established individual contributor—never wants another promotion. Picture this person in your mind; write down their attributes.

   a. Now imagine this person showed up on your doorstep and you HAD to put them on your team tomorrow. How would you leverage their differences? What new opportunities would you be able to explore?

3. Now imagine you ARE this new person; the maverick of the group. How would you feel about the team environment? Would you feel safe? Valued? After three months, when the honeymoon starts to wear off, are you still going to want to work on this team? Will you feel like your unique traits are worth something to the team, or will you be tempted to start hiding some of your differences in order to fit in?
   a. To get more insight, ask yourself: how often do you ask your team how you can make their day better? How often do you sincerely ask for their input, and incorporate it?

Chapter 3.

Put responsibility on the teams

I want to talk about ramps. You'd think I would have nothing but good things to say about ramps, right? But the thing is, as someone who uses ramps a lot, I can tell you, one only works for me if I can get in the building once I'm at the door.

In the 1990s, when North America started to recognize the value of accessibility standards and practices, it might have seemed like a great moment for people with disabilities. Finally, companies were putting ramps and elevators as part of general construction. The problem was, decision makers often didn't consult with people with disabilities or those who work with us before implementing the changes.

So you'd see these milestone moments where new codes are like 'of course we need ramps and elevators,' but the buildings would have no automatic door openers.

People affected by proposed changes, and those closest to them should be consulted. Always.

At first door openers simply often weren't included. It was demoralizing and frustrating to those who finally felt they had access. Essentially, it was a failed mission.

Then Personal Support Workers (PSWs) and people with disabilities started making their voices heard, saying there's no door openers. And that is when codes started changing in ways that made sense.

Imagine how much more efficient and successful the accessibility rollout would've been if the people affected by the changes had been part of the consultation and building process in the first place.

My experience as a manager goes back to university, when I hired and trained my own team of support workers. If someone called in sick and there was no replacement, I'd be left stranded. But at the same time, I had things to do! I couldn't let my time get taken up with scheduling issues.

So I handed it off to my support workers. I told them they were a team, and their goal was to make sure I got the support I needed, regardless of which worker covered which shift. I made it their responsibility to get their shifts covered by talking to each other. Once they made an arrangement, they'd let me know about it. This freed me up to focus on my schoolwork and the rest of my life, and it freed them to get a day off here or there when they needed it, just by talking to each other.

Do you see what I did there?

1. I gave them freedom to make decisions. Their "support rail" or limiting parameter was this: someone needed to show up in the morning to help me. They got to decide who.

2. By letting them handle scheduling issues instead of me, I supported them and treated them like human beings. It's much easier to ask a colleague to cover your shift than to ask your boss for a day off. So they got to call in sick, or arrange a trip, or do whatever else they needed without having to explain it to me. And I got my time back.

It seemed like such a basic idea to me at the time. I was just so much better off giving up control over the scheduling. I did have some fear of letting go; what if no one showed up? But fear can either motivate you or limit you.

If you let fear limit you, you're accepting that you're just not going to be able to try something new that could be good for everyone. I wasn't willing to accept that. I wanted to try the new thing and see how it went.

When the team leads the leader

Interestingly enough, I didn't initially adopt the same style when I started managing people in the business world. Looking back, I was too busy trying to be an individual contributor and have all the answers. I thought I needed to because I was The Manager. My team was building websites, but I was taking on everything myself and working late nights and in a silo to make sure everything got done. Basically, I was making the mistakes I now coach other leaders not to make. I was more worried about managing the work instead of managing the people.

It was one of my personal support workers who showed me the light. Her name was Dorothy and she always carried one of those organizers with her, remember those back in the day?

She was with me one day and noticed me struggling, and she said very simply: "That's not the way you lead us."

Well, no, I explained, this is different. She made me wonder though, was it different?

So I started leading the way I led my personal support team. I set goals and let my team figure out how to reach them. And you know what. They took it on. We did it.

Later, Dorothy said, "this should be natural to you," and only then did I start seeing the broader picture of leadership and management independently.

What happens when companies miss the bigger 'picture'

Now I want to take you back to 1975, when Kodak invented the first digital camera. 'Wait, what?' you ask. 'I was still using film in 1975. Digital cameras didn't go mainstream until the early 2000s.'

True. Stay with me, because that's the point.

What happened between 1975, when Kodak researchers invented a digital camera and when you bought your first one, was a bewildering missed opportunity to capitalize on the birth of digital photography. Back in the 1970s, film was still thriving and the new digital technology seemed clunky and slow, and Kodak decided to shelve it rather than launch it and let it compete with film sales. It was a costly decision. Within a few years, Japanese manufacturers including Canon, Sony, and Fuji developed their own digital cameras. Kodak sued and got patent royalties, but their patent expired in 2007 and Kodak filed for bankruptcy in 2012.

The response time between the trigger and response—opportunity and investment—took too long.

Blockbuster suffered a similar fate during its home movie wars with Netflix.

Originally, Netflix was an online DVD rental service; you chose your movie on the website and it came in the mail. Then you mailed it back. It was basically a digital version of Blockbuster. In 2000, Netflix was losing money and its CEO, Reed Hastings, wanted to partner with Blockbuster. They would run the online brand, and Blockbuster would promote Netflix in their brick-and-mortar stores, Hastings said.

Needless to say, Blockbuster passed. Its leadership saw Netflix as a "niche business."

Also, Blockbuster bosses didn't like that it meant no lucrative late fees. The video rental company made a lot of money off late fees—and its customers absolutely hated it. Any store clerk could have told Blockbuster how much their customers resented those made-up due dates and the fat profits Blockbuster made off them. Returning your rental to the store was inconvenient and that's why people were always late. But Blockbuster was blinded by dollar signs.

That's why, about a year before Hastings flew out to pitch Blockbuster, he made a simple change to his business model. He made it subscription-based. Instead of paying per rental, you paid a low monthly fee with no shipping, *no late fees*. You could only have one or two DVDs out at a time—but you could keep them as long as you wanted, and to return them, you just dropped them in the mail. To make this work, Hastings had to change his website and adjust his software.

Netflix's business started to improve right away. It was still struggling in 2000, but before long it started disrupting the entire industry. Netflix's secret? It kept decision-making functions close to the product teams that were working with their customers.

Blockbuster may not have cared that its customers found late-fee policies inconvenient and downright predatory, but Netflix did. The new company created a business model to work *with* customers instead of penalizing them. They put the customer first.

John Antico, the Blockbuster CEO, eventually saw the problem. In 2004, he eliminated late fees and launched Blockbuster Online. But the changes cost money and his people questioned his leadership. He was fired within a year.

By 2010, Blockbuster was bankrupt.

The point is, listen to your customers and employees—the ones on the front lines, not the ones scheming for your job. They have their fingers on the pulse of the customer base.

Keep employees at every level in the loop

When I tell people to push decision-making to the front line, they tend to get nervous. 'Trust those people with my business? But they don't know the first thing about our strategy!'

Well, there's your first problem. Start there.

The employees who actually do the work need to be engaged in the business. They need to understand the vision and how their role fits. They need to know your *why*.

What's the point? Why does the company exist? Why does their job exist?

You may be thinking, that's what onboarding is for. You do that as soon as you bring them in the door. You probably also tell them they'll have autonomy. You've been looking for someone just like them. Their voice will be heard.

Then you show them to their cubicle and tell them how to do their job.

Organizations fail when they tell employees exactly *what* to build instead of explaining *why* they're building it. The ideas around what to build should be coming from the people who will actually build it.

The top-down approach, where managers deliver decisions that have already been made, creates teams that are good at listening and delivering, but not good at sensing the client's needs and responding for themselves. Instead, build your diverse team and then actually use it! Instead of giving team members the solution you want them to build, give them the specifics of your problem. Then ask, how can we best solve this problem for this individual? The team may identify different good angles, based on their different perspectives. That gives you *options,* instead of the top-down approach that says there can be only one solution: the one management picked.

That's why two-way communication is important. Management doesn't know everything, and can't, unless the bottom is talking to them. Teams need to be able to provide feedback on the tools they need. It's important for management to understand the cost of not providing that tool. And you only get that intel from a team that feels confident speaking up.

To be clear, "tools" doesn't just mean the latest software. Some of the things employees need, to be their best, revolve around their personal life—such as flexible work hours to support childcare arrangements. The better the environment you can create for your employees to articulate their pain points and have that pain minimized as much as reasonably possible, the more your employees can give to the organization.

Now, most managers wouldn't refuse to accommodate childcare arrangements if asked. That's unreasonable! But too often, we expect that we can just sit back and wait for our employees to tell us everything they need. If they don't, they can only blame themselves, right?

That's not good enough. As leaders, it is our responsibility to ask our employees:

- What are they struggling with?
- What's their biggest challenge of coming into work—of being at work?
- What's the number one thing they worry about while at work?
- How can you make sure they leave work and be able to be their best selves at home?

An organizational chart should be looked at as a communication highway, not a management highway.

One of the things that makes change so hard in legacy organizations is the elaborate organizational structure they've put in place. There are so many layers and levels that will be thrown off by any change. Your organizational chart should have a clear structure but should be flat enough that decisions can flow quickly from the top to the bottom.

Change is no longer a project with a start and finish. It's continuous. So your organization needs to be built for responsiveness. Responsibilities should be outlined and delegated just enough that the areas of responsibility and channels of communication are clear. An org chart should be useful for someone asking themselves, "Who needs to know about this?" rather than the older hierarchical model, which was designed to answer the question, "Who has the power to make this decision?"

You shouldn't need to ask yourself how many layers of power to go through to get your answer, because your job as a leader is to create an environment where growth and decision-making are happening at the bottom; an environment where Agile can thrive.

Set the tone from the top—include clear safety margins

In today's way of building things, the manager communicates the organization's purpose and educates, inspires, informs and creates the environment around the teams. This structure allows managers to focus on what only they can do: set the tone from the top.

As facilitators, you have to set the standard for your organization's culture and create the safety margins for your employees. Like those guardrails I mentioned earlier, you need to have fallbacks in place so your employees can feel comfortable to fail, knowing it won't blow up the business.

This is why it's important to set out recognized safety margins for the team: These can start out small. Then, as you begin to build trust, you give wider gateways.

For example, make it clear employees can take chances as long as they don't affect the schedule more than two weeks, or as long as the company would be able to recover from the worst-case scenario. Margins give employees space to figure out for themselves.

These type of recognized safety margins—that indicate how far people can travel outside the box—help give teams the ability to react to situations in the moment, without needing to travel to get a decision on the right response.

Traditionally, in the corporate world, the general vibe has been that employees should leave their home lives at home. But that was in the days when you *could* separate work and home.

In recent years, even before the COVID-19 pandemic, it has been increasingly difficult to separate life and work. Personal and professional worlds—blurring since the dawn of smartphones and VPN—have merged since COVID-19 forced millions of employees across the globe to work from home.

Employees are reachable almost 24-7. But as the pandemic has also shown, that level of being "on" leads to burnout. Management needs to recognize that if we are going to continue making demands of our employees when they're supposed to be at home, then it's only fair to factor their home life into what we ask of them at work.

Otherwise, we create a perpetual weight on our employees that grows exponentially. If we're not meeting their needs at work, if we just constantly *take*, they won't be able to meet their other commitments. They'll skip out on the fantasy football league. They'll scale back their volunteering. They'll be less available for their families. That weight will keep piling on, aided by guilt and fear, until despite all their skills and abilities, it exceeds what they can hold up.

Do you remember my story in Chapter 2, about working in a building where I couldn't use the washroom? I didn't ask for help until I felt safe, because it took so long for me to get a job and so many jobs before that one hadn't been accessible to me in my wheelchair.

Finally, when I landed that one, I didn't even think to survey the facilities and make sure I would be accommodated. It wasn't until I went to go to the bathroom on my first day, that I realized I couldn't. So I decided to hold it. But even that decision didn't fix my biggest problem. I was working longer than eight hours a day; the tasks I was hired for didn't align well with my abilities, because my hands are really affected by cerebral palsy, and I couldn't do all the paperwork.

When technology gave me a computer, I could put it to work reusing files, building databases and—once the Internet came in—even get people to register for their own classes. And I was a damn hero for it.

I became the facilitator of the system. When I was able to do all my work in a third of the day, instead of a day and a half, they started using me in other groups, looking at the way they worked (or not) with a new eye, and figuring out how they could use these new tools to become faster and more efficient.

In the 90s, change was seen as a bad thing. You were supposed to plan well enough to eliminate change. Change meant you didn't do your job. But at that time, the world didn't change as rapidly as it does today. These days, change isn't something you avoid. It's necessary to survive.

In fact, historically that's always been the case. Before the Industrial Revolution, people's lives were routinely derailed by weather systems, illnesses and political turmoil. In some parts of the world, it's still that way. But in developed countries, everything goes smoothly. So we started to think of change as a one-time event. We designed buildings to last 50 or 100 years. Some did—but not without adapting.

Beyond Version 2.0

Traditionally in software development, manufacturers would build a shiny new bit of software, and then every year after that they'd make it a little better or more efficient. Version 2.0 and beyond.

But now that's not fast enough. Customers are changing every day. People switch from phone providers to Internet providers in a heartbeat. Families who'd been contemplating a switch to online groceries for a decade started instantly once the pandemic hit.

Think of what a leap of faith online shopping was in the early 2000s. Were you worried someone would steal your credit card number? Now it's common to autosave credit cards in our browsers. As our confidence grows in technology, we accept its rapid changes. We live in a global economy with more competition and more options than ever before. People who can work or buy services anywhere, we have to understand what they want better because they are gaining influence every day.

It used to be what customers see is what they get. Now, as products compete with ideas and innovation around the world, it's more like what customers want is what they get.

The question for management is, do you want them to get it from your company or somewhere else?

Disrupt yourself

To keep up in this fast-paced market, you need to disrupt yourself. That means finding new ways and new markets. It means prioritizing flexibility over efficiency.

And as leaders, that means being flexible in how you manage employees. Every individual is different, even when you need to manage them within the same system. Each of us have different challenges and needs and things that keep us up at night. As a leader, we've got to really know our people to create an environment of safety, so they don't sweat the small stuff or put a lot of energy just into not being wrong. We need to know where they are and meet them where they're at. We need to proactively ask, "What can I do to help you do your best work?"

This is a departure for many of us who have been conditioned to see the boss as someone employees have to impress with their best work, rather than someone who needs to help those employees do their best work.

Putting responsibility on the team requires what's known as **servant leadership**. And it doesn't stop at asking questions. Servant leaders might allow their teams to design their own workspaces. Heck, they might jump in and help rearrange furniture. They might pitch in with the grunt work on a project with a tight timeline.

The great leaders we've had, the ones we think the most highly of, all did this. We remember them because they went above and beyond, taking things on that weren't related to their work, just to make our lives better. If you help create an environment where you're the first to offer help and people are comfortable asking for help, your team will thrive.

Behaving as a **servant leader** who jumps in where needed, creates rapport and builds morale for the whole team is how you create that magical "we're all in it together" energy that gets people so engaged.

Constantly re-recruit your employees

To keep employees engaged, you have to re-recruit them. Instead of handing them the specifications for the solution you've decided on, give them the problem earlier and ask for their perspective. Instead of telling them to "tone it down" and learn to fit into your corporate culture, celebrate the fact that they bring something new. Remind them what makes them so good at what they do.

Consider the role of a grocery store cashier. These types of jobs are often incredibly over-supervised. Breaks are timed, workstations are monitored. Clerks are literally told where to stand and what to say. We have all worked jobs like this or know someone who has.

Imagine if a store trained its cashiers to make sure the customer has the best possible experience. Imagine if cashiers were encouraged to ask for tools they need, like an ergonomic mat or a handheld scan gun for large items. Imagine if the floor manager were to hold a weekly meeting where cashiers shared best practices and tips with each other and forwarded any customer feedback they received. How different would that store feel when you walk in? How would that sense of agency and control cause those cashiers to behave differently?

After several months of working remotely during the COVID-19 pandemic of 2020, some banks surveyed employees to learn what has changed for them since moving online. The overwhelming request from employees was for a way to emulate in-person board meetings. This request resulted in digital whiteboards, which changed the dynamic and motivated the employees.

If instead, you as a leader focus on maintaining control of the process, you will inevitably limit what the organization can achieve as a whole.

If employees are worried about being "wrong" they can't learn and get better. If wrong isn't acceptable at your company, learning and getting better will never happen.

Be curious. Ask questions. With leader silence, the bottom will go silent.

When I was at my first job, I could have benefited from a manager who had enough bandwidth to notice I never used the washroom.

If the one I had even asked specifically about my workplace needs and whether I needed anything to be at my best, I probably would have told him. I'm no martyr. But at that time, he hadn't created a safe environment for me, so I didn't get to contribute fully to the organization.

If you want to empower employees to speak up, you have to start by speaking up yourself. The model for career advancement used to be, you progress because you're better at the job; you have way more experience or skills. And, once you had advanced, it was your team's job to make you look good. That's not the case anymore. The modern manager doesn't know everything, but they know how to learn everything. And now, your role is to raise up everyone around you on your team. Being the smartest isn't the goal anymore. It's not even desirable. Instead, you need to be able to create an environment where your whole team pushes or pulls together to elevate everyone around you.

Notice that I said, "everyone around you." As a modern leader, you are part of your team—not apart from it. That's not to say you don't have a role to play; you have something important to offer. You have niche knowledge about the organization's history; what was tried in the past and how they got where they are today. In any discussion, you can lend perspective. You're the custodian of the organizational vision, and you're the best at pulling together intel that the team brings you, framing it for them so it makes sense.

A few years ago, I led an engineering group. At the time, it had been forever since I worked on software development. I couldn't lend specific insight into the technology solutions. But I was great at creating an environment where the team could collaboratively solve their problems and build great code a supportive environment where problems could be exposed and differences of opinions could be collected, so we could really consider all available approaches. Then I'd give the reins to the team to decide which experiment they wanted to try first. That way, instead of jumping on the first viable idea we came across, we could settle definitively on the best one.

Too often, we jump at the first answer instead of considering all of the possibilities, the different ways we could experiment, and what success might even look like.

As a manager, following these guidelines will help you deliver the best outcomes, consistently:

- Practice sitting with the problem, rather than rushing to a solution.
- Learn to draw on your diverse team for their perspectives: ask good questions, challenge them to come up with different ideas. Those ideas will help orient you.
- Create a team that is comfortable being skeptical of its ideas and doesn't feel compelled to defend itself to you
- Allow for failure: let team members know "reasonable" failure is to be expected and you will all get together to learn from every failure.

But there's also a time for silence.

I once took part in a meeting through Microsoft Teams with maybe dozens of colleagues. The subject was the Surface trackpad and at one point it became clear at least some of the participants in the meeting were not aware I was present.

In a response to a question about functionality, someone posited, "What would Dave say?"

I had trouble unmuting myself and in that instance someone else responded, "He'd say to make it customizable."

Which was exactly right.

When it comes to accessibility in products, customization is key because an individual's abilities are just that: individual. Other people with cerebral palsy might have completely different needs from, say, touchpad functionality than I would.

It was awesome to hear that my message was getting through, especially after I'd started at Microsoft with so much anxiety over whether or not the role was for me.

Sometimes leaders can take a step back and just listen, finding out what's being said around the campfire, so to speak. See how the culture is responding, find out what is being distilled.

MID-CHAPTER EXERCISE

The Echo Test—Leading Without Saying a Word

Purpose:

To help you assess and cultivate the kind of leadership where your values, vision and priorities are carried forward by others—even when you're not present.

Background:

You know you're making an impact when your ideas, values, or strategies are echoed by others, especially in spaces where you're not directly involved. This kind of leadership goes beyond charisma or authority—it's about embedding purpose into the culture.

Step 1: Choose Your Core Message

Ask yourself:

- What is the one principle, cause, or strategic focus I want to be known for championing?
- What do I want people to say about me when I'm *not* in the room?

Write it down in a single sentence.

Example: "Dave would remind us that this accessibility thing is important."

Step 2: Map Your Influence Network

Think of the people you interact with regularly:

- Your team
- Peers in cross-functional meetings
- Stakeholders
- Mentors or mentees

Next to each name, ask:

- Have I clearly communicated my core message to them?
- Have I created opportunities for them to carry it forward?

Highlight any gaps.

Step 3: Look for the Echo

Over the next two weeks, observe meetings, conversations, or decisions you're not directly driving. Look for:

- Someone referencing your ideas without you prompting them
- Others advocating for your vision, priorities, or cause
- Feedback or comments that sound like your influence, even indirectly

Write down at least **three moments** where your message showed up without your direct input.

Step 4: Create Amplifiers

Pick 2–3 people in your influence network. Ask yourself:

- How can I equip or empower them to carry the message further?

- What stories, data, or frameworks can I share that will stick?

Schedule a conversation or 1:1 to share your thinking and listen for how they interpret it. Leadership is about passing the torch, not guarding it.

Step 5: Reflect and Refine

At the end of the month, come back to your original statement and ask:

- Is this still the message I want to echo?
- How has it evolved?
- What new signals have I seen that my leadership is resonating?

Get comfortable with failure

In order to get the best from your teams, you must get comfortable with failure. But first, put guardrails in place. Create clear parameters for when teams can proceed without you, and when they need to bring you in. This is their playing field, and you're their coach. Except you're a good coach, the kind that doesn't need to call plays because their players are so well-trained they already know how to react. You run the practices and condition your people so they can use their best skills, ability and judgement in the moment.

In his memoir, *Eleven Rings: The Soul of Success,* NBA basketball coach Phil Jackson explains he didn't coach players, he coached the environment. He knew when to call time-outs, when to let the players relax and when to give them a rousing pep talk. He had spent enough time conditioning the players, so he could focus on the flow of the game. He knew they could perform if he created a supportive environment for them.

Great company leaders manage the flow of the culture and keep an eye on what's going on in the business. Your goal is to manage the environment so your teams, who are actually building the products and services, can use their skills, develop, and learn.

Change is constant, so get on friendly terms with it

I've spent a lot of time describing what leaders *should* do. I can just as easily say what they *shouldn't* do—that is, break down the projects into tasks and tell everyone exactly what to do. You shouldn't do that, because you shouldn't have to; you should have already provided a vision to your employees that gives them the guidance they need.

Like Jackson, your focus needs to be bigger. If you're busy barking orders and making sure everybody follows them, you won't have the resources left to sense what your team needs. And that's your real job as a leader—to figure out what they need.

Remember Netflix and their subscription model? That came about because someone told Hastings about Blockbuster customers hating late fees. He listened and asked his team to help him solve that problem.

If you can get your team and organization comfortable with the idea that change is a constant, it will create a lower-stakes environment for your ideas. You just need a solution for right now—not forever. That will allow people to be braver and more outspoken, which will open the door for other ideas—even contradictory ideas. If you don't need an idea to work forever, it doesn't throw a wrench in things if you find a better one.

Think of a TV series. Every week there's a problem and the heroes solve it, but the show's not done. Next week there's another problem. Project work and change is cyclical like that; you solve one problem, and there's another one. We can be heroes over and over again.

Most people don't come up with a great idea in one fell swoop. Even if you have a divine vision or sudden breakthrough, there's still a lot of work to do to develop your solution. The best ideas are the ones that can be built upon and changed. They get beat up and thrown around, turned upside down, tossed away and picked back up. That's how you know an idea is ready, tested and thought through.

But if you've ever tried to test someone else's idea in a meeting, I bet you weren't met with enthusiasm. We're not always comfortable with people picking apart our ideas, even if they're trying to be helpful. It feels like getting attacked because there's so much at stake. We don't want to fail. We don't want to look bad. But what if we weren't worried about those things? How much more could we get done? If you lower the stakes, the tension will become lower too.

We know that teams think better than individuals, because biases get filtered out. But groupthink can become a bias; if everyone just wants to get along, you end up with a consensus culture where it's considered rude to point out a problem. A good team knows to challenge each other and take an idea and build upon it, because every great idea has to be built.

Think of American football. Every so often, the quarterback will throw the ball all the way into the end zone and it's a huge victory. But that's not how the majority of goals happen. The real grit in the game comes from the team moving the ball first down after first down, 10 yards, 10 yards or maybe only three yards. It's really taking the downs as they come. But if you do enough downs, statistically speaking, your big opportunity will come and you'll be able to pull off something ridiculous.

The more ideas you bring up and shoot down, the more times you fall and get back up, the more likely it is that you'll happen upon that lucky break and get to run all the way.

Let teams fall and get back up as they work their way to goal achievement.

If your team is going to perform at their best and give you their great ideas, they need to understand what success looks like. Success doesn't mean you never fail along the way. Rather, you should have a supportive environment in place where everyone falls and everyone helps each other back up.

Just like my walk on the beach. The falls weren't failures. They were the way to the water.

Encourage employees to stay in learning mode. Lead by example.

I love learning new things because it's one of the spaces where I feel free to be wrong.

I'm just in learning-mode, after all. I find that one of the most productive phases occurs while I'm trying to master that steep learning curve. Once that curve plateaus and I feel like I've "mastered" the subject, I feel pressure to be good at it all the time. If I'm in a meeting and they're talking about product management or software development—areas where I have a lot of experience—I get nervous about sharing my input. Because, what if I'm wrong? That could be embarrassing. That nervousness hampers my creativity and innovation. If I don't feel free to be completely wrong, I can't completely learn.

If I feel that way, it's important for me to recognize my team members do too. And if I want the best for them, I should encourage them to continuously stay in learning mode.

Being an expert in a domain is great. Continuously learning new expertise is even better! You need to believe every success is a new beginning that leads into the next challenge—the one you haven't mastered yet. And remind your employees that.

Reassure them that even though you appreciate their expertise, you expect them to tackle and learn new things as part of their role. That every accomplishment leads to a new start, where they are allowed to be wrong again.

If you embrace that mindset, innovation will flourish, because you'll be in a growth mindset versus a fixed mindset. The minute you get the mentality that you can't be wrong, you lose your creativity. You lose that childlike ability to see old problems with new eyes.

Freshman mindset

I went away for university, like many young people. I did it because I wanted that chance to start fresh with new friends with new perspectives and idealisms. When you surround yourself with completely new people, you get a new perspective on yourself. Suddenly, you feel free to move past your old limitations and try things you never imagined. Your abilities haven't changed. What's changed is the people around you and how they see you, which can sometimes overwhelm who you want to be.

When you see change and learning as continuous instead of single events, you become *built* for change. And when you're built to change, you can't become obsolete.

How does an organization set up guardrails and make failure affordable? How do you find the balance between safety and freedom? It's trial and error accompanied by ongoing support.

Conveniently, most companies already have a few supports that they provide to their employees. These include things like a three-year business strategy, their corporate vision, and some institutional knowledge about customer needs that are being met with existing products.

Cross-functional teams

I recommend building cross-functional teams so employees can get to know how the different functions work. This will give them a practical understanding of your business' end-to-end operations. Once they have that, your corporation's "why" will make so much more sense to them. They'll have the whole picture.

At the most basic level, what I'm saying here is that modern organizations and leaders need to let their people work holistically. That means two things: 1) creating an environment where employees feel safe and empowered to bring their whole selves to work, and 2) giving your employees the knowledge and vision to see their organization holistically.

Because every problem, whether it's business, technology, or team, is really a people problem. To solve any people problem, you start by looking at the environment. If you create an environment that supports and empowers *human beings*—not a corporate ideal of an employee, but the specific employees who actually work for you—you will find many people problems resolve themselves or improve significantly.

Unfortunately, too many organizations have a way of making employees feel that whatever makes them human is exactly what the company *doesn't* want.

Microsoft hasn't only allowed me to be my authentic self in the workplace; the company's put me on blast, broadcasting my way of life in both internal and external marketing campaigns.

But that doesn't mean there weren't bumps in the road. When I was filming that video for the Surface Adaptive Kit, some folks in Marketing and Legal were trying to control my narrative and make it as palatable as possible. They were giving me a bunch of script notes, and I didn't know how to be the perfect disabled person. I just knew how to be a real one.

That's when Panos Panay, Chief Product Officer at Microsoft, gave Marketing a directive: "This is Dave's story. Let him tell it how he wants to."

Allowing me to present my story as best I could, warts and all, helped manifest a spot that more gracefully connected with real people, real users of Microsoft products.

Let your team know you value their so-called 'soft skills'

I don't like the term "soft skills." There's nothing soft about the skills often labeled soft. In fact, they're absolutely critical. Technical skills are a tool for individuals. I can only do one person's job with them. If I leave an organization, my technical skills leave too. But being a good listener, having empathy and building trust allow me to build skills in others. They let me connect, collaborate and get the most out of my team. The effect of my soft skills builds over the years, as the people I helped help others. Soft skills have a multiplier effect, letting me improve many people's jobs with one hour of my time.

That's how strongly they can innovate.

That's why essential skills—as I will now refer to them—make great leaders. And great leaders build the leaders of the future.

Remember I said I once led a software development team, years after I stopped working in software development? My team knew the technology. I didn't. That would have been awkward in a traditional company, where I should have known the tech well enough to dictate their every decision. In this organization, I knew nothing about their technology, but I knew how to get them collaborating as a team, brainstorming, challenging each other's ideas, and building experiments to figure out which idea was best. I knew how to get the best out of them. I spent all my time trying to build an environment that they'd want to bring their best selves to, so they could come up with the best solutions.

In a traditional organization, I would use extensive product knowledge as a stick to wield. But if you have to use the carrot-and-stick method, with nothing but your authority to back them up, you're not doing your job right. It's important to create an environment where people understand the rational—the *why*—behind the decisions. If people believe in the cause, they don't need incentive; they are empathetic to how this solves their problem. You should never need to do that. Instead, you should have enough personal knowledge of your team, enough rapport with them, that you can influence and angle them to perform at their best without needing to be asked.

This might open up new possibilities for them, and increase the likelihood you'll learn something new that you would have never expected out of your team members.

My superiors at Microsoft, for example, gave me the opportunity to lead a wider team, integrating employees working in ergonomics. This was important to push accessibility forward because, again, Microsoft didn't want to assume a disposition that disabled people should be thrilled we're acknowledging them through our products. We wanted to provide them with the best *choice* of accessibility products, which meant the tools couldn't just work, they had to be *comfortable* to use.

But my oversight of ergonomics also meant I was affecting the production of items used by able-bodied people, too. Panos, the Chief Product Officer at Microsoft, once said to me, "I know with you I get accessibility for free, but you can do so much more." He realized Microsoft wasn't unlocking my full potential.

It reminded me of when I went to university. The school administrators put me in the accessibility residence, dormitories that were easier for me to navigate. But being relegated to that building meant I was alienated from much of the general population. Why couldn't I be friends with people who played football, too?

After a couple years, I moved into a regular dorm, which is the kind of setting where I swim better, if I may.

When I am expected to engage with something that's particularly niche, I get bored quickly. It's usually too limited in terms of scope. And working strictly in accessibility at Microsoft for a time made me think, "Did I make myself too small?"

I like the complexity of disability, but I also like helping humans of all shapes, sizes, and abilities achieve more. So when my team got wider, I got excited. It helped me expand my horizons and broaden possibilities for the products being developed under my supervision—not to mention the learning opportunities for my team members as well.

How to navigate old-school managers

I'll give you a warning. If your organization is fairly traditional and you start trying to bring these ideas and practices in, you will likely run into the old-school management mindset. Many employees came up with that ideal of leadership in their minds, and it still affects them today. They saw the smartest people getting rewarded and rising to the top. And what did they do once they were at the top? They told everyone else what to do, of course.

Things have changed so much since then. Today, leaders who don't listen are eventually surrounded by people with nothing to say.

Now, emotional experience is considered valuable, because it helps develop critical skills. But when it comes to technical experience, in an organization that's built for change, 20 years of software development experience isn't that much more valuable than five or 10 years. Technology changes at a constant pace, so all of our old skills are now constantly expiring and we have to develop new ones. You may have seen the articles laughing at a job ad for requiring 10 years of experience in a technology that's only been around for five years? Well, having 20 continuous years of experience in C++ is just as ridiculous today.

If you have middle managers on your team and you start emphasizing the essential skills we talked about before, listening, empathizing and responding, they may feel a little confused. They got where they are under the old system, where they were rewarded for being the smartest or the best at fitting in to the culture. These are their glory years—when they can kick back and let their teams do all the work, then step in and accept another promotion!

That approach isn't good enough anymore. As a leader today, you should be present and hands-on when your people need you, but invisible when they're shining.

We need critical skills today more than ever, because emotions move people more than facts and data. Facts and data are inputs, and we have more of them than we could ever need.

How disability has helped me demonstrate the value of innovation

People who live with disabilities understand the value of creativity better than most. I've had to become innovative just to get around the world as a person with a disability. I see problems differently because I had to. Time and again, I've had to figure out a solution that would work for me. And much of the time, my solution didn't exist yet, so I had no frame of reference for it. I had to dream it, try it and refine it until it worked.

I've also had to learn how to delegate responsibility, a skill that many technically skilled employees never learn. When you're the smartest and the best, why would you let someone else do the work? That's why so many managers cling to control. But that's not how I learned to manage.

My limitations have always been very visible, and let me tell you something: I can let my wheelchair limit me, or I can use it to move forward. It depends how I choose to use it.

Technology not only helps people with disabilities live a fuller life, it bridges the gap between their lives and those of able-bodied people. That's why I went to Microsoft.

But tools specifically designed for people with disabilities aren't the only ones that can have such an impact. I don't think I'd be married if there was no such thing as online dating.

The profile I created on the platform through which Kelly and I met included the fact that I had cerebral palsy. And I did post photos of myself, but the inherent functionality of the website allowed Kelly and I to get to know each other as unique individuals, with physical characteristics being a secondary concern. What I mean is, we messaged a lot and then spoke on the phone, for hours. And as I mentioned before, I'm a good talker, and listener.

The platform also opened up the possibility that we could meet at all, which otherwise was slim because she lived 40 minutes away from the city I was living in at the time.

When Kelly first met me, she wound up being a little overwhelmed by the prospect of having to provide me with so much assistance getting around. After our first real date she told me she didn't think she could move forward with our relationship for that reason. I was disappointed—OK, *devastated*—but there wasn't anything I could do about it. I wished her well.

Three days later, though, she called me, told me she missed me and that she wanted to give the romance a shot. So our connection, made initially possible by a digital platform, text messages and emails, before the phone came into play, carried us past challenges that might otherwise have proven non-starters.

Horizontal career paths

Career paths that we used to think of as "vertical" (get a promotion and hopefully a raise as you stay in the same profession/department but move up its hierarchy) are today more and more "horizontal" (get good at a trade or skill; go to the department next door and get good at that trade or skill, etc.).

This horizontal/lateral career path is more popular today because of people's curiosity about many career disciplines.

More and more, we need to think in terms of learning paths, not career paths. The idea of a career path can pigeonhole people; "Oh, he's not a leader. He's a career software developer."

Facebook Chief Operating Officer Sheryl Sandberg once famously described career development as a jungle gym, not a ladder. This is such an important idea to bring to your employees. Often when people move companies, it's because they don't see their future in the organization—or they do, and they feel limited by it. Horizontal growth opens up all these new doors. Now they can make lateral moves, enter a whole new world, and develop a whole secondary set of skills, without having to leave your company.

And the more your employees can do this horizontal growth, the more well-rounded, creative and versatile they become. They learn how to continuously learn and change and grow, like that cycle that I mentioned earlier. They get good at change, good at learning. Exactly the kind of people who can step into that role and fill that need that never existed before in your company, but now it does.

Employees who have been through multiple departments get exposure to all the different ways that people do things *within your company*. They're not going to bring in ideas that won't work for your organization. They might bring in ideas that do work, but that were rolled out in a different department, so your team didn't know about it. This will help break down internal silos and even help your teams become more self-aware.

A newcomer from another team will have a new perspective and be in the right position to say, "Hey, you guys kind of make life difficult for this other department when you do this."

Have you ever daydreamed about taking a course or program, only to realize that your organization won't fund it because it's not in your "development plan?" That means, they don't see a business case for it right now. But what about five years from now? That always bothered me. I'd feel like, *how dare you dictate how I grow? I want to be the one deciding how I grow.* Trust me, if one of your employees thinks a certain skill set is necessary and relevant, they may be seeing something you're not seeing.

I remember years ago when my colleagues were told they couldn't learn a new technology, because "we don't use that here." When companies behave this way, they put blinders on themselves. First of all, there's no way you can know that your company *won't* use that technology within a few years. But more importantly, learning is always valuable. Maybe learning this technology would give me a different perspective on the technology we *do* use now, and I'd see a way to do things better.

Horizontal growth also solves the problem of employees having to tell us exactly what they want. They might not know! Often, we give people raises and promotions thinking that will make them happy, and they accept because that's what they've been told success looks like. But if an employee isn't driven towards a certain goal, why give up that flexibility? After all, we don't know what kind of workforce we will need in the future. So why limit people to their current roles? If we keep their careers open-ended, we won't have to play catch-up later.

How to grow your employees internally

First, build learning into the timelines of your projects and deliverables. If you don't make time for it, it won't happen. Make sure there's slack in the system to let learning happen, whatever it might be, instead of insisting on a tight timeline. Let them set the timeline, and tell them to add time to research, ask questions, explore options, and even upskill. If they find an online course that will give them the skills to build a better product, don't you want them to do that now and not after the product is built?

We're used to thinking of roles in terms of rank and responsibility. But in reality, every role is comprised of skills. Instead of looking at building a career, why not build a skills inventory to show what your organization has and needs, so people can map their own careers out.

Again, this will push the decision-making onto the actual employees, rather than limiting them to the organization's vision for them.

For example, consider running a career fair annually. It will give people a chance to learn what each new group is up to and help define their learning path.

Your recruitment teams can even do this as a precursor to going to external career fairs. That way, you're re-recruiting your own employees by showing them other areas where they might be needed. Nothing is better than if we re-recruit them to a new team or project, because then we're growing them, but we also get better messaging externally—*Look at the growth opportunities we offer our employees!*

This approach will keep your people happy, empowered, and growing. Talent will stay at your organization and continue to grow, ripening for you to pick whenever you need it. We don't have to abandon people because we have a new need and they don't fill it anymore. Instead, we can re-recruit from other groups within the business, empowering our employees to move around without having to leave the company. If we create the conditions where anything is possible, it usually is.

Today, no one is building a better mousetrap. Instead, we're all just making different mousetraps. That's why we need diversity; to tell us where that unfilled need is that we can tap into. Allowing employees to drive lateral growth in their careers means helping keep their perspectives wide and therefore more diverse.

Exercises

1. Once you've decided to set up this empowering environment for your employees, how will you figure out how much control to give them?

2. In your own organization, how will you expand the authority of your employees?
    a. Consider tracking the decisions you make each day, and how many you delegated. Because as a leader, you should strive make the minimum decisions necessary to allow your team to make progress.
3. How will you encourage your employees to come to you?
    a. Track how many times employees are coming outside of your one-on-ones, because that means you're approachable, and if you're approachable, you're valuable.
    b. Once you've got an idea of how approachable you are, track how those conversations went. How many times did you give an answer, and how many times did you ask them a question and send them away to think about it?

Chapter 4.

Change for growth, one imperfect step at a time

For many years, the very idea of dating was so difficult for me that I didn't even try.

As a teenager, I had crushes and imagined having romantic dinners with girls I was attracted to, but every time I thought about actually making it happen, so many barriers came to mind that I'd get overwhelmed and dismiss the idea. I worried about even trying to explain my disability to a new person—something I knew I'd have to do.

It's an easy thing to say, 'I have Cerebral Palsy,' but it's very different to get people to understand what that really means. And even if a girl I was talking to could accept me, I felt like her parents would probably have objections. Parents want their daughter to be cared for—not have to care for someone else.

My hometown of Windsor, Ontario, is an industrial city. A lot of our parents worked at car factories, and I worried that some people felt if I couldn't do physical labour, what could I offer?

And those imagined challenges in my mind were just the first obstacles. What about the logistics of dating?

Let's say I met someone by talking over the phone, we hit it off and decided to go on a date, 'then what?' I wondered. I would need my date to walk me or drive me. God forbid, my mom drove us!

To get from the car to the restaurant, I would need help. Even if I could bring my power wheelchair, I'd still need assistance getting in and out of it. We'd have to plan ahead to make sure the restaurant was accessible, and the menu had food I could eat without asking for it to be cut up.

Of course, I'd have to stop drinking liquids at a certain time the day before, so I wouldn't need the washroom. I mean, asking someone to help you pee on the first date may be considered a bit too forward in many cultures.

At the end of the meal, I'd want to take my wallet out to pay—and do it quickly, before my date thought I was trying to hold out so she'd cover the bill. That's not an easy maneuver for me, but I wouldn't want to ask for my date to help get my wallet out; she might feel bad.

There were just too many unknowns. It was exhausting and demoralizing. And I wondered, even if I went to the effort of figuring out each detail, would it be worth it? What if there was no one out there who could accept me? My efforts would always be for nothing.

Despite all these reservations, I couldn't stop thinking about dating. And I'd imagine every potential date as The One. My future wife. We'd go on another date, and maybe she'd ask me back to her house. We'd build a serious relationship and then get married. I was thinking all the way to the finish line, but I didn't know how to even take my first step.

So instead, I'd hang out with my guy friends. For them, the path to the perfect date seemed to unroll at their feet. But I didn't have a ready-made path. If I wanted one, I had to make my own.

But back then, I preferred to stay in my comfort zone. It seemed less painful than taking a risk to try to find something better. In other words, I gave up.

It was only when my friends, who'd always been there for me, started getting into serious relationships, marrying and starting families, that the risk became more worth it to me. They were less available for me. I realized I had a choice: find somebody or learn to be a hermit.

So I decided to ease myself into the world of dating. Fortunately, I had friends who were girls. I would go to dinner with them to get used to being in a one-on-one setting and to go through the details. I learned how to ask them to help me up, walk me to the door, fold my wheelchair into the trunk of their car.

Those "pilot" dates opened my eyes to the fact that any date I went on wouldn't just be a new experience for me. It would be completely new for the woman, too. I've always had a mentality of having to adapt to my circumstances, but now, I would need to help someone else adapt to my differences, become confident and comfortable with me so it would be worth all this effort.

Suddenly, I realized I didn't have to find The One. That was way too much pressure—not just for me, but for her, too. I just had to have a meal with someone. We'd take it from there.

It was that mindset change, looking at how my dating project would affect the other person, that started me on the path to success in dating. And it occurred because I experimented and took small risks.

Organizational leaders can have similar success by considering how their teams and all members can contribute and will be affected by the desired change. And then work to support them.

Leaders often focus on the organizational change they are creating with their new projects and initiatives. But change impacts everyone in the organization. It's critical to think through how you are going to help others manage the change, rather than just focus on rolling out your new initiative. It's not enough to just be the one in control, making all the decisions. You also need to be willing, to make yourself vulnerable and to see the project through the eyes and experiences of your team.

In Chapter 1, I discussed some behaviors leaders need to adapt to the rapid pace of change in the 21st century. But beyond those new behaviors, today's leaders need a full-scale mental transformation. You need to cultivate a new mindset where you are constantly learning, shifting and growing, and you're comfortable in that state.

In my industry, we call that an agile mindset. It's about becoming comfortable with being vulnerable and open to new perspectives. That can present itself in different ways, but no matter what you are trying to achieve, if you can develop that flexible, open, agile mindset, you will have a recipe for leadership success.

You make yourself vulnerable by leaving your comfort zone in the hopes of learning something new.

When I finally went on my first date, that mindset allowed me to give myself permission to be comfortable and show my vulnerability. That layer of vulnerability is what really makes you effective.

For me, that meant asking my date to move my fork so I could reach it, asking for a straw for my drink and asking to have my food cut up.

But vulnerability isn't just a nice-to-have skill for me. It's my reality. If my date turned out to be a reckless person, or someone who wanted to harm me, there might not be a lot I could do. I was really putting myself in their hands.

Luckily, with the rise of online dating, I had a chance to screen people and get to know them before going out. But this was before most people had cell phones—I couldn't just text my friends if the date was going badly.

I'll never forget one of my conversations with my wife when we first met. We were talking about going on a date, and I asked if my wheelchair would fit in her trunk. She said, "Oh yeah, my trunk is so big you could fit two dead bodies in there." We hadn't even gone on our first date!

Superhero Syndrome

Sometimes I would try to fend off the vulnerability by overcompensating. I'd try to make myself sound like

Superman. It was just insecurity and of course, it wouldn't last. I can stop a speeding bullet, but can you please cut up my meat?

But it's not only me who feels tempted to overcompensate at times. Everybody has a disability, so we're all doing this to some extent or another. Maybe you get into a new group of people, and now you're trying too hard to fit in. That just makes it worse. Or you're not trying hard enough, because you want to seem cool and confident. That doesn't work either.

Once I started going on real dates, I had a new challenge. If my date didn't call back or said it wasn't going to work out, I didn't know if it was because of me, or my disability. At that time, I was trying to separate the two.

Eventually, I had to learn that my disability is part of who I am. It can't be separated. So when a date didn't work out, I had to just accept it didn't work out, and keep trying. The more I did it, the more comfortable I got with my disability. The more comfortable I got with my disability, the more comfortable other people could get with my disability.

What this means for leaders today is that when you're trying to embrace possibilities, you can't be on the defensive. It's so easy to focus on developing those coping mechanisms, like my overcompensating. But as long as you're only *coping* with change, you'll always feel one second away from drowning.

Back to my walk on the beach. The best thing to focus on while you're going through change is how far you've already come. Think back to where you started out and appreciate the journey. That's how I went from thinking I hadn't walked far enough, to realizing that I had come farther than I ever expected.

The One

So when I finally met my wife, I was ready... almost.

We were matched on an online dating site called Lava Life. I sent her a virtual carrot, which is how you showed interest. Soon we moved our online conversations to Messenger. I told her about my cerebral palsy and the limitations that come with it.

Her third question out the gate was, "Can you have sex?" That might seem awkward or forward, but for us it broke the ice and set a foundation for honesty. She gave me permission to be open. I replied, "I could tell you, but I would rather show you."

Our first date was great. For me. We went to Kelseys for appetizers. It was overwhelming for Kelly. Despite my pre-planning and our foundation of honesty, it was still a lot for her to come pick me up, put my chair in the trunk, drive me in the car and get me home.

That experience made my disability real to her. It felt like too much.

"I don't know if I can do this," she said later. "I've always wanted somebody to support me, not someone I have to support."

There was some irony in her saying that to me. I was in my 30s and she was in her 30s, and I had lived away from home while she never had. So I mentioned that. And said I was actually pretty practiced at living independently. I just did it differently. But she wasn't convinced. I'm sure she was doing the exact same self-defeating thinking I had about dating; that it was all or nothing, marriage material on the first date or never seeing them again.

At the end of our conversation, she said, "Well, maybe we can still be friends." I tried to put a good face on it, I said, "You know, to be honest, I have enough friends. If this isn't going to work out, I wish you all the best." Truthfully, I was devastated. But so was she. Two days later, she emailed me and said, "I'd like to give it another shot."

So we did. The next time, it wasn't such a big deal to pick me up at my house, or walk me to the car, or fold my wheelchair into the trunk.

In other words, once the scary new thing becomes ordinary—and it always does—it will open the way for even more new possibilities. The agile approach is about embracing that trial-and-error period, staying in it until what was new is now normal.

Reconsidering your assumptions about change

I hope you'll remember this story and it will help you find new ways to think about change. I have come to recognize three dangerous assumptions people often hold about change:

1. Change means a tradeoff—you might end up with something better, but you will definitely lose something, even if it's only the status quo.
2. Change is for other people; that, whatever big project or transformation you have in mind, it will require change for others. But not you.
3. Change is a choice: one you can opt out of.

Maybe it's easy for me to say. For me, change has always been a necessity. I had to learn different ways to get dressed than other people, different ways to leave the house and different ways to accomplish tasks at work. I couldn't expect the world to change for me, or others to adjust their ways of doing things. I had to be the change.

When the usual way of doing things doesn't work, you have to choose whether to step out of the game or keep playing. If you want to stay in the game, change is the number one rule. You need to change the usual way of doing things to a different way that works for you.

Change doesn't just even the playing field, it raises the bar.

So for those who think change is a tradeoff, ask yourself, what are you *really* giving up? In the same sense, if your organization is failing or struggling, what do you have to lose?

And to those who think you can opt out. No. The fact is, change is a necessity for everyone. Your choices are to engage with the world as it changes, or to fail. It's not true that change always requires you to lose something, but it is often true that if you don't adapt and change with the world, you will lose.

The growth mindset

Having a growth mindset encourages leaders to inspect and adapt, which means we don't have all the answers figured out up front. Instead, we develop a minimal solution and try it.

In my day-to-day work this is part of what's known as an agile framework. Working in this framework, we understand that we'll likely fail the first time we change, but we'll also quickly apply the learnings from that failure to the very next attempt. At the same time, we think about what really motivates the people around us: autonomy, purpose, and mastery. This has to be approached as a discipline.

All around you in your organization, different value systems and hierarchies can distract you. You have to constantly reaffirm with yourself that this is about progress, not perfection. You have to set up training cadences to go through the exercises and read up on it, much like yoga or an exercise routine.

Having a growth mindset is about starting with a mental state that can benefit anyone. You have to develop this mental state in yourself, through a commitment. That part was easy for me; I had little choice but to commit to change. For most people, it has to become a mental discipline, constantly driving your focus towards intrinsic motivation, a growth mindset, towards delivering value, towards continuous improvement.

If you're thinking of being agile as a soft skill, something easy, then you're already missing the mark. It's not simple, and it's not like a fashion trend that you can adopt for a season and then move on. It's not a solution; it's a discipline, and using that discipline is a craft just like any other.

Status quo syndrome

Organizations who resist change are in effect disabling themselves. The disability may not be physical, like mine, but they do have one. It's called the status quo.

Most organizations cope with this disability by sticking with what they know. They might not be happy, but they don't want to sacrifice their stock price today to go for something that could be better in the future. Nor do they think they can afford the learning-stage period, where time and resources are spent developing a new model. What if the tradeoff isn't worth it? It's a case of "better the devil you know."

That mindset is crippling. You might as well have one hand tied behind your back.

This is one of the areas where organizations struggle. When they do commit (or are forced to agree) to a massive change, leaders often demand employees perform at the same level as usual. It's a ridiculous expectation. No matter how skilled you might be at dancing, if I ask you to take up oil painting, you're going to struggle. Even transferable skills will take time to translate as employees take on new processes.

Mistakes will be made

Last chapter, we talked about how it's critical for leadership to allow room and space for learning. When teams are learning, they are not the expert. Same goes for leaders. But remember, to learn something new, you need to be open. You need to be vulnerable and make mistakes. As you make those mistakes, you'll have the opportunity to learn from them. But in so doing, it's important to accept giving up a bit of productivity, because you're not as efficient in a new process as you were in the old comfortable one.

Without an agile mindset that embraces change and promotes trying new things in small steps, that point of lost productivity is the danger point for organizations.

It's tempting to say, "See, this isn't working. Let's just go back to the old way."

It looks like change is failing if you measure success based on your old status quo, which was the end state of the practices your organization spent years building. You can't compare something that's been tried, tested and true, to a new complete way of doing things.

(Expert) Identity Crisis

It's not only a fear of failure; some people are also very invested in that identity of being the expert.

Kids don't have any insecurity about what they don't know. They want to learn new things. They openly ask questions. But as we get older, we start to hide behind our credentials as the expert in specific areas.

You can focus on valuing expertise and sticking to the status quo, or you can give yourself a chance to grow new capacity by becoming a novice again. But you have to pick one, because they're mutually exclusive.

I think professional growth requires us to get out from behind our Expert Identity. That's why when I kick off a workshop, I'll ask participants to tell me something interesting about themselves that I can't find on LinkedIn.

One critical struggle for organizations is knowing when and how to reinvent themselves. As the world changes around you, there comes a point where you need to transform. But that means redefining yourself, coming up with new practices, and worst of all, giving up control. It means sacrificing your current expertise in order to learn to do something new and better. That sacrifice is so hard for leaders and organizations to make, particularly for publicly traded companies. Because stockholders don't like it.

If you work in the financial industry, you know the pressure of the quarterly cycle. Your organization's stock price takes precedence over long-term sustainability or positioning for future growth. Because if investors lose faith in us, we won't have a future.

It's a classic example of the urgent hijacking the important.

In today's market, organizations need to stay in learning mode as much as they can. It's the only way you can find the most innovative solutions for your customers. And you don't just walk into the sunset once the project is complete. You continue to track product performance and collect customer feedback through the life of the product.

This approach is almost the opposite of traditional project management. Typically, you get a large scope for a project that will take over a year, and you build everything into the product that was requested in the initial scope. But with that approach, you don't know what will happen once you release it and a customer starts using it. Even with market research, there are still so many unknowns, and so much money at stake.

It's easy to find yourselves in a bubble, listening only to the data sources that affirm what you already believe. You can develop a theory of what your customers want and need that has nothing to do with your actual customers or their real needs. If you don't spend time observing your customer in their natural habitat, you won't get to see the opportunities or problems you can solve for them that they'd be willing to pay for.

Building relationships by constant communication with customers

Using an agile approach that encourages small, consistent changes, companies are focused on product into the customer's hands as quickly and cheaply as possible *and then keep learning.*

What that does is get you in close contact with the world of your customers. Instead of spending all of your time building and executing, you spend a significant portion of time listening, learning and researching. That's how you can build real relationships with your clients that will inspire loyalty.

This model allows you to enter into a dialogue with the customer. Instead of speculating internally as to how the market will receive your next product, you turn a sample size of your customers into beta testers. They try the product for you, their behaviors change, and they tell you what they like or didn't like.

By tracking real product usage, you can find those unknowns that are so scary for a product manager. The customer might use your product in ways you never even predicted. That's not a surprise. It's a "learning."

That's modern leadership: sensing and responding. Listening. Collecting information, then experimenting based on that. Leaders today can feel threatened by change, feeling they're stuck in an unwinnable life-or-death situation, but that's because they're trying to restore what they have instead of inventing what they need.

When looking for areas to reduce cost, organizations often cut items that seem non-critical, such as training and customer visits/travel. But I argue that when things get tough, that's when you need to learn the most. Whether you do or not, the market is going to change around you.

Adopt Different Perspectives

Working in accessibility at Microsoft has made me a more empathetic listener than I already was. And when you're employed at a company that makes products designed to appeal and satisfy an extremely broad range of consumers, you must listen closely and try to adopt the perspectives of others.

This has shown up in my responsibilities as a product maker and accessibility champion. Sometimes those two roles cannot be reconciled to the satisfaction of everybody.

But I do my best—which is all I can ask of myself.

This requires detailed discussions about design options versus budget and returns on investment. And both sides of my professional mind have to remain open to hard realities, i.e., tradeoffs. Oftentimes taking a stance comes down to me asking myself, "Is it better to have a slightly lesser product than the more optimized one I envision on the market or no product at all?"

Early on at Microsoft, I spoke with an accessibility advocate from inside the company who ran through all the things she felt our Surface devices were lacking. Her wish list was, like, 50 alterations deep, and she asked me, "Do think they should all be fixed?"

"Are you asking Dave the product maker or Dave the disabled person?" I asked.

Dave with cerebral palsy wished all of them could be added to the devices. But the Dave that considered cost knew it wouldn't be feasible.

Balancing those perspectives has allowed me to go to Design, for example, and advocate for product upgrades, but from a place of more deeply grounded business reasoning, recognizing what's truly possible and vital to the customer base. In a world of cross-functional teams—the kind of workplace in which I like to live—this is highly applicable and key to functionality.

Catastrophizing

It's also important to stop thinking of failure as a terminal state. We often fear *being a failure*, as though our mistake is now going to be our legacy. And it leads our fears to compound one another. If we don't know how to achieve our goal, it amplifies our fear that we might fail.

Our fear of failure is multiplied by the fact that we don't distinguish between failing while trying something new, and being a failure. It's called catastrophizing.

That's exactly what I was doing when I scared myself out of trying to date. I imagined obstacles for myself: In my mind, I let the possibility of an "inaccessible restaurant" become a catastrophe, when in fact we would obviously just go to another restaurant. I created a negative framework that held me in place, scared to take a single step for fear I would fail forever.

But change is ongoing and infinite. You're never going to not be doing it. Best to get comfortable with it. If you fall along the way, it's not a broken leg. It's a skinned knee. It's an opportunity to inspect how far you came and what you've learned. Then you adapt for the next step.

Change at the hiring level

When organizations do try to prepare for transformation, they often make one of the biggest possible mistakes right off the bat by hiring someone with years of business domain experience in the area where the organization has always excelled, rather than hiring someone who has the right behaviours and learning mindset to be able to take the organization where they need to go next.

The right leader is one who can accept the direction the organization needs to go, and start filling in the blanks on what needs to change in order to get there. Much like me as I began to date: I knew I wanted to start spending time with girls instead of groups of friends, and needed to figure out what to do differently to get there.

When you become a leader, you often pick up where the previous leader left off. Your company often just expects you to extend the same business model that's already established… for decades.

Over time, you lose your ability to embrace change. In addition, any major business transformation will have significant costs. That means leaders are reluctant to change, because they could be wrong. They're entering a place of uncertainty.

A company doesn't just jump from A to Z when it comes to change. It's organic; it unfolds. And it doesn't happen in a vacuum.

Scrum and agile values are focus, openness, respect, courage, and commitment. We use those values to drive ourselves through that empirical process I've described.

An Agile mindset says, we're all in this together and it impacts each of us differently.

I'm sure our dates impacted the women I went out with differently.

EGO, EGO, EGO

Often when I encounter resistance to change in an organization's leaders, it's ego-driven.

After everything a leader has achieved, and everything they know, they can gain a sense of identity that becomes an ego. It protects us from being wrong; it doesn't allow us to be vulnerable.

And it affects how we value our customers. An ego-driven leader can become condescending towards customers, like we know what's best for them. We can dismiss their concerns, as though there's something wrong with them for not knowing our product is perfect.

This is the huge risk of the expert mindset; what makes it a disability. You're told you're the expert. It builds your ego. You want to live up to that perception; you stake your pride on it. Now you have something to lose, so anxiety comes in.

But if all you're focused on is protecting your ego, preserving your identity as the expert and dealing with your anxiety, you're not customer-focused anymore. You're me-focused.

You're not fulfilling your true mission in the company.

The reason most companies value expertise is that it makes things more efficient. Experts should be able to do the job faster and better than others.

And of course a healthy amount of ego is linked to drive and purpose. The key is, it needs to be combined with vulnerability and discovery to create a powerful force that drives innovation.

In an agile framework, leaders become expert in entering what's called the "discovery phase." They become comfortable with the rough terrain between idea and implementation.

That doesn't mean skills don't matter; of course they do! You will always need to be good at your craft. But you'll need to build skills like listening, connecting to others, and setting a vision.

With Agile, learning is not temporary. It's a mindset. So there will always be a new horizon to conquer; a new level of depth or breadth you can attain.

This is where vulnerability comes into play again. To return to my dating story, I was good at being a friend, but I needed to expand that into the opposite sex. How do I go one-on-one? How can I be vulnerable with my disability, when no guy my age wanted to seem vulnerable in front of the opposite sex?

I had to approach it as a discipline. I had to purposely get comfortable with being vulnerable by putting myself in situations where I would feel that way, over and over again. I needed to build supporting behaviors like resilience, to help push through the vulnerability and the fear of getting hurt.

One way to manage this fear of getting hurt is to remind yourself that whatever learning phase you're entering, you're only a guest, not a resident.

Whatever discomfort may seem to be coming up on the horizon, it's just part of the cycle. What's new today is tomorrow's routine. That routine frees up capacity for you to explore another new horizon, and then you get comfortable again, and find another new area. As a leader, you are always a visitor, never a resident. For better or worse, whether you're experiencing success or failure, it's always temporary.

Pick your battles

Once I started to get more comfortable at Microsoft, I began speaking up a little more. My initial reluctance to do so, I think, paid off. If you scream your head off all the time about everything, people will grow desensitized to it. But when you wait to persuade folks strongly, doing so in moments when you're the most passionate about something, chances are better your perspective will be embraced.

My kind of coming-out party at Microsoft arrived when I was trying to get a haptic trackpad integrated into a laptop for accessibility purposes. Design and I went through multiple rounds where I pled my case. I was told each instance that they weren't ready yet. "We want to make sure it can work for all devices, not just one model of laptop," they said.

It happened so often I began to feel self-conscious about repeatedly advocating for the change. I emailed a colleague and asked, "Am I being an asshole by constantly bringing this up?" He typed back, "You're the only one here who can challenge accessibility on a product and *not* be an asshole. You're asking the things that should be asked."

Reaching peak frustration, I finally said in another meeting with Design, "This is Microsoft. We innovate. If we're going to innovate, let's just put it in one device, learn, and then integrate it across all devices. But if we're never going to do it, let's stop wasting time even discussing it."

After the meeting, the Senior VP of Product told me, "I want more of *that* Dave, not the version who's reluctant to speak up."

The haptic trackpad went through for a laptop model and it worked well—a huge jumping-off point for the technology.

Since then, I don't think I've ever had to push for anything that hard. But I might have to again one day, and doing it so seldomly will make it count when I turn up the heat.

Change your party story

Do you have a story or two that you like to tell at parties? The story of how you met your partner, or a funny story that always lightens the mood?

Imagine you're at a party, and instead of telling one of your best stories that you've told time and time again, you start telling a new story that doesn't even have an ending.

Who would do that? It's embarrassing. People won't get it.

That's what it can feel like to change organizational approaches, like giving up something successful and established, to move to something that is uncertain, could very well be embarrassing, and that people might not even understand.

It's normal to find change difficult to accept, let alone embrace. It's scary for a reason—it might not end well!

But tolerating change is about forgiving yourself for failures, even if those failures undermine your identity as an expert. We seem to develop this fear sometimes that if our expertise gets undermined, then our identity goes too. Like we're nothing without our knowledge. It's like me when I didn't feel I could date: it left me feeling almost hollow inside.

A new route to work

In contrast, remember how exciting it is when you see something new for the first time. Like trying a new route to work, for example, and seeing new neighborhoods you'd never seen before. If we take the same path to work each day, we don't get to experience new surroundings and we don't find a new path. We cheat ourselves by limiting that exposure to new things because we want to get efficient. Especially because the reason we want to get efficient, to learn shortcuts, to take less time or energy to do something, is to make room for the new experiences.

I hate to see a leader or organization so stuck in their identity as an expert, but not using that extra capacity for something. You have an ability to push yourself from novice to expertise. Why don't you give yourself a chance to do that again? Why are you focusing on maintaining what you have instead of growing something additional?

One of the best ways to build resilience and tolerance for change is by creating a vision for where you want to end up. Individual leaders don't have a monopoly on their visions; others may know better how to reach it or may see that it needs to be adjusted.

That's where you will need some vulnerability, to allow others to adjust your vision. It can be difficult to allow your vision to change, especially if the change is coming from others. But if you keep going with a bad vision, that's just a recipe for failure.

Sometimes, you've got to be brave to set your vision. Look at Amazon. It bet against the American shopping mall. That was a pretty bold vision. But it also led up to it; they started with books and CDs and expanded from there. As people got more comfortable with buying online, Amazon offered more and more products. And they adjusted the model bit by bit.

Exercises

1. Think about some places in your personal or professional life where the thought of change gives you a

sinking feeling. Write down some of those resistance points and times where change has been suggested or you've thought you might have to change and found it really scary to consider.

a. When's the last time you took on a new hobby, tried to learn a new skill, read a different type of book outside your favorite genres, or listened to a completely different kind of music? Learned a new computer program, a new type of technology, contacted people with a new medium? Are there any technological trends that came and went, and you just never got on board? Last time you rearranged your desk? Because even small changes in your environment/routine can prime your brain for growth.

2. Ask yourself: what routines do you have, daily weekly, monthly or annual, that you would be willing to change?

a. Try to walk yourself through reimagining one of these comfortable routines. Notice how it feels. Is it uncomfortable to imagine doing this routine differently? Do you notice a sense of inertia pulling you to reject the new way?

Examples—cooking a new dish to add to the "rotation"? If you're going to adopt an agile mindset at work, to do it successfully you'll need to be able to allow it to bleed over into your personal life. Because that's how you learn to get comfortable with change/growth/trying new things. You may even start to find it exciting. When we start something new, our ego drives us to want to be good at it. Resilience comes in when we say, "I know I'm not going to be good at first, but I'll keep trying until I am." Vulnerability comes in to allow us to not be an expert.

Chapter 5.

Leaders as teachers, teachers as learners

Years ago, while working at a financial institute, I needed a new program manager for my team. The position I wanted to fill was essentially a manager responsible for helping a team find its most efficient processes and then coaching team members to work together and become successful at using those processes.

I started off trying to get top talent with expertise in the field of course, but when I saw I wasn't getting candidates I'd hoped for, I decided to reframe my search.

Instead of focusing on *skill set*, I wondered, what would happen if I tried focusing on the *behaviours* I couldn't live without, and built from that? With this new mindset, I recalled a candidate I had interviewed previously for a different role in a separate department. I knew from the way he articulated his answers, that I was sitting across from someone with the potential to be one of the best managers I could ever have the privilege of working with.

The only problem? He didn't know much about coaching teams.

What he did have was a ridiculously high emotional quotient (EQ). And it was clear, he had the right personality and the drive to learn quickly. That's a character trait. It's tough to train people to develop traits and behaviours. On the other hand, training someone to develop skills and knowledge is relatively easy with the right candidate.

I could tell that if I invested in this person, gave him the knowledge framework and the skills to start with, he would absorb everything and work hard to get himself to the level he needed to be at.

When I recalled that interview, I knew I had an opportunity—a responsibility, even—to develop top talent. So I made him an offer.

He said yes and joined a team that was dysfunctional because of a general lack of communication, discipline and direction. He brought much-needed change. Within weeks, I'd hear laughter and chatter coming from their area and the team started delivering frequently and with high quality.

But one of his first challenges was to help the team get through a significant backlog that I knew was about double the amount of work the team could be expected to handle according to their deadline.

The new manager was optimistic. He asked his team and when they told him they could handle the workload, he vouched for them. "The team feels they can get it done."

As a leader, it's not always your job to give people the solution and tell them what to do. Sometimes, it's helping them frame the problem and giving them the responsibility to solve it. Have them come to you and others for advice, but give them the control to make their decision. You have to allow your team to take these risks.

I was skeptical. But I let him try. If I'd stopped him there, and told him to get a new plan, he only would have reconsidered because I told him to. But there was also the possibility he could really do it—and then I'd be the one with something to learn.

He didn't. They missed the deadline. And my new product manager—who would typically walk into my office with a big smile, full of energy—looked defeated when he came in to tell me the bad news. Here's how that went:

Him: "I screwed up."

Me: "Okay."

Him: "The team didn't finish the sprint. I screwed up."

Me: "I'm disabled, not deaf. What are you going to do about it?"

Him: "Well. . I think I'll notify the stakeholders, work with the team to find out how much work is left, and see how creatively we can do it to reduce the extra time needed."

He did what he said. They found the efficiencies and a creative solution.

Interestingly, a few weeks later I was in the same predicament. This time, I was managing a couple of teams, and we had a deadline coming up. Considering the data and listening to the enthusiasm of my team, I thought to myself, "They're gonna do it. I can inspire them."

They didn't do it. So I called the junior product manager into my office. I wanted to show him that failure is okay, and what better way to do that than to make myself vulnerable by showing him that I had failed?

Here's how that conversation went:

Me: "You know, I have a lot of experience. I should have known we weren't going to make it. I screwed up."

**silence as I continue beating myself up.*

Me again: "I can't believe this ... I screwed up."

Him: "Dave. I'm not deaf. I heard you the first time. What are you going to do about it?"

In that moment, I went from being a leader to being a follower again. And I was learning from someone I had just mentored.

Once again, I think back to my wobbly walk on the beach and my friends who were there time and again to help me up to take another step.

That walk, like those few weeks at work both serve as reminders that success isn't a straight line without any bobbles.

It's often about falling and getting back up. Sometimes you're the one falling, and sometimes you're the one lending a hand to help someone back up.

It takes courage to allow your employees to see that you need help sometimes, too. And courage to accept that help from them.

Yet, more often than not, I see leaders skipping regularly scheduled one-on-one meetings that can help build relationships with their employees and develop an understanding around growth goals. It's a trap that we all fall into—too busy putting out fires to work on relationship building. But the only way to move your organization forward is to put your trust in other people and be vulnerable in front of them.

Leading as a teacher and as a student

In my own career, working as a trainer has helped me become a better leader. I learned to empathize with people who struggle to understand a concept. And I can confirm that the better you become as an educator, the better you become as a communicator.

As a teacher, I help my employees learn to navigate change, gain new skills, communicate and engage with people. I believe leaders have a responsibility to grow their employees by exposing them to new challenges and areas of knowledge.

Many of my employees have a business or computer science background. Sometimes, the area they most need to learn about is human behaviour, so I challenge them to see the people and environment around them with new eyes.

On the other hand, I've also learned a lot from my teams along the way. Sometimes team members take the teacher role and I become the student. As a student, I am open to my employees' perspectives. I know when they are confident enough to disagree with me, there is an opportunity to learn. The notion that higher titles mean leaders have nothing to learn from their teams is a dangerous one.

Some managers feel weak without enforcing an authoritative top-down structure. But true weakness is in that hierarchical mindset. Vulnerability gives you real strength: Knowledge is empowering and learning can only happen if you admit you don't know something.

No matter how inexperienced they are, any employee has something to teach you. It could come from their personal background, their diverse perspective or a unique work experience. So, when you see that an employee is thinking differently about a situation than you do, ask why. Give them a chance to explain. Find out what you may have missed. This will add to your toolkit for next time.

Always take the opportunity to be a beginner. It takes courage to not have it all figured out, and not try to get it perfect first time. But being an avid beginner allows you to learn. As long as you're constantly putting yourself in new situations, you keep your sense of curiosity alive. That allows you to learn new skills before you actually know what you need them for.

Training vs. learning

Training is isolated to specific concepts and theories that you study, with discrete learning outcomes. You train to get specific skills and knowledge. Your training experience is often removed from your day-to-day work. When I design a course, my students are taken out of their environment and I teach them new tools that aren't related to their everyday work. My hope is they learn enough to evolve their work and use those tools on the job.

In contrast, learning is more holistic; you can learn from any experience and you can learn from anyone, whether they intend to teach you or not. All you need is the ability to observe what's happening around you with a curious mindset.

But good learners do more than just observe situations and behaviours around them. They can also discern what's meaningful about that pattern; they see the ramifications and how to apply the pattern for next time. As a leader, your job is to allow space for that process to happen and create flexibility for the plan to change based on those learnings, rather than cling to the existing plan. As your team is learning, you're continually making new plans to incorporate those learnings.

In other words, expand your focus beyond output. Your team isn't only about output. They also need inputs, and one of those is development.

Let the smart people you hire teach you

We all know that highly intelligent employees can present a challenge to their bosses. You can tell by talking to some candidates, that they're the type to ask questions and shake things up just by their nature.

If you're operating a hierarchical model of authority, that can feel threatening.

But if you're truly focused on what will be best for your team and the organization, you'll be drawn to those people.

As Steve Jobs said, "It doesn't make sense to hire smart people and tell them what to do. We hire smart people so they can tell us what to do." Hire smart people now and learn from them in the future.

This is more easily achievable if you meet people where they are. I get further ahead by working with people I trust, and accepting their knowledge base, viewpoints, and opinions.

I was late to the accessibility product making game, but I've been able to get ahead, I believe, because I fully engage with people who bring a variety of perspectives, materializing from different personal and professional journeys. This is true for disabled people I've encountered, as well as able-bodied individuals who may or may not need convincing that accessibility is a huge plus in business. This has helped me develop diverse, high-performing teams that have implemented agility-forward changes or developed products. It can apply to any team-building process, really.

Meeting people where they are helps you bring them along on the team's journey. You can't assume everybody's right with you in lockstep. But when they come around, you can help them grow as a part of your team, so it's important to remember: meeting people where they are doesn't mean they can't also change as they move forward.

How do you achieve this somewhat abstract goal? Imagine you have access to a time machine. Go back and visit your own self at an earlier part in your career. What were some of your stances on relevant professional issues? Have they changed? I bet they have.

Now you can more easily empathize with your colleagues, recognizing they're entitled to be who they are right now. This facilitates what I call "Capital T Leadership." When you're managing a team, your goal is to drive them collectively upward, like the stem of a "T." But you also have to effectively engage with your peers across the leadership tier of your company, existing literally on the same line as you in a corporate framework. Accepting every person's authentic self, even if they create friction on the path of achievement toward your goals, allows you to manage all these different personalities more evenly.

Return on learning

As leaders in today's rapidly changing world of business, it's just as important to look at your **return on learning (ROL)** as it is to look at your return on investments (ROI).

ROI shows you how you're doing through a financial lens. But ROL tells you how well you *can* do in terms of growth. Corporate growth isn't just about money. Just because you can't measure intangibles like learning doesn't mean they're not important. Sales figures offer a snapshot of where you're at *right now*, but that's only relevant right now. ROL tells you where you can go and what you can do.

I'm hoping that this book and the examples it included have made the case for looking at your company's ROL.

We need to look at trends, and growth is a trend. One way to measure it might be by looking at how far you've come over time. In that sense, growth would be a lagging indicator of learning.

Make it real. Give learning a budget.

Offering learning opportunities for your employees starts with your budget. Fight for a budget that will allow you to constantly grow your team. I'm not talking about training once every few years. I'm talking about continual investment that might extend to areas irrelevant to your employees' current jobs.

Take a page from Google. In their 2004 Initial Public Offering (IPO) letter, Google founders Larry Page and Sergey Brin, famously wrote: "We encourage our employees, in addition to their regular projects, to spend 20% of their time working on what they think will most benefit Google. This empowers them to be more creative and innovative. Many of our significant advances have happened in this manner." Google News, Gmail and AdSense all developed out of "20% projects."

It's clear Google doesn't just have a loose vision for their employees to keep learning, the organization embedded learning into the roles and responsibilities of employees. It's part of a plan.

You too need a plan for each employee who reports to you. They should help create it. Then make it official.

To develop a plan, recall the exercise from Chapter 1 about creating a skills inventory and looking at your gaps. Referring to this inventory, figure out how you can grow your employees together. If you can get peers teaching each other, you get the added benefit that they're also learning teaching skills. And we know that leaders need to be teachers. So, by training teachers, you're training your organization's next generation of leaders.

The power of diversity

When you hire people with diverse backgrounds and worldviews, you get to learn from a variety of cultures, work functions, and perspectives. This is a huge opportunity for you as a leader to learn from your employees.

Unfortunately, this opportunity is often overlooked. Leaders who hire a diversity of candidates often spend all their energy trying to manage that person into their team's groupthink culture.

Instead of trying to get your diverse hires to think your way, try thinking their way.

During the 1980s and 90s women were entering the workforce in large numbers and rising through the ranks. It was the age of the power pantsuit, remember? Women were being told that if they wanted to work with men, they had to work like men.

As a result, organizations and society in general lost out on the learning opportunity right in front of them—to learn from their female employees' unique perspectives.

The point is that while it may seem like a perfect fit for your team is someone who thinks like the people you already have on it, that's actually not the best choice for the organization.

An outlier with a diverse opinion might be the force you need to go through the pain of getting to the right answer, instead of accepting the first one because everyone in the meeting agreed.

Study after study has shown that diversity boosts innovation and helps organizations succeed.

That's why it's essential to hire for the role, not to replace a person

Hiring to replace a person means you're looking to get the same kind of person you had before. But that's not how you evolve and grow.

When you're hiring, instead of remembering that last person you had and trying to replicate them, try boiling the job description down to the most essential skills. Then seek out fresh perspectives.

Once you've brought on a "diverse" team member, you need to deliberately support their differences. Because once they enter the role and try to fulfill it, there's a risk that they'll settle into it and lose that change factor you were after. As I discussed in

Chapter 2, the only way to keep that cognitive diversity is to make sure everyone feels valued and able to bring their true selves to work.

Do you remember the TV-series called the A-Team? I loved them as a kid. They each had their own individual weirdness, but together they just work. When hiring, don't focus on conserving what you had before. Envision your own A-Team, where each member is weird in the perfect way to complement the others. Ask yourself, what's the next member your A-Team needs? What's the next skillset you need to add in?

If you're looking at each candidate in terms of 1) essential skills and 2) that extra diversity of experience or perspective, you'll be able to find someone who adds value to your team.

Transferable skills are found where you least expect them

For example, imagine you need to hire a salesperson. Your last team member was a career sales professional with 10 years of experience. Now, you have a candidate who was an actor and improv artist, and is looking to switch careers into sales.

On paper, it looks like a terrible fit. But in real life, that hire could be a powerful step forward for your team as a whole. They don't have industry experience, but they can adapt to conversations on the fly, they're really good at interacting with people, and can read the room, which is a skill set they could teach the rest of your team.

Inclusivity: More than just a buzzword

These days when you talk about "diversity", many companies still think about age, race and gender. And diversifying in those areas is an essential part of growing an organization. But it's also important to consider inclusivity in an enterprise sense, where individuals feel connected to the entire organization, not just their department.

Sometimes you have to condition your team to become more inclusive and open-minded. By positioning yourself as a student, seeking to learn as much as you can from the diversity around you, you signal to your team that this is important to you. They will start to own responsibility to make sure people are included and new hires can settle in without being asked to sacrifice the things that make them unique. If your employees can make room for that new person and their new ideas, they're creating breeding grounds for innovation.

This isn't something you can mandate. You can only model that learning mindset, and hope your employees will take their tone from you. If you're always encouraging your people to contribute their unique perspectives, it becomes something they can do with confidence.

When people are comfortable, they do their best thinking. So as the leader, you need to create that safe space, where people feel they can bring out their quirks and be welcomed.

How can you get an organization, top to bottom, to welcome inclusion? Show stakeholders how *everybody* wins with diverse teams—which is what actually happens, so this isn't that hard. Still, I'll provide an applicable story that can serve as a metaphorical example.

When I was in fifth grade, I once went many consecutive days not really participating in gym class at school. I mostly just sat on the sidelines and watched my classmates play basketball. Finally, our teacher pulled a group of kids together and brought me onto the court with them. Back then, I could still move around a little, crouched on my knees. Together, we hatched a new version of basketball with a modified scoring system.

Dribbling and shooting was out of the question for me. Catching the ball, while difficult, was something I could do if it was passed to me at just the right speed and angle. So in this new game, the team I was on essentially had the advantage of two goal options: the basket and my arms. If the ball either went through the hoop or into my arms, it was two points.

In this scenario, my teammates were motivated to pass the ball with a light touch and accuracy, which took practice, while I worked to improve my ball-receiving ability. I also had to learn how to get open, just like any other player.

We might have still technically lost some games, but my teammates and I got even more out of phys ed, which was the real win.

If this story is a strong enough metaphor, you can see how organizational inclusion is more than a nice to have—it's an opportunity for employees to challenge themselves and broaden their perspectives. This will lead to stronger output.

Don't punish failure

If people on your team feel that their projects carry a risk of career suicide if they screw up, they'll never feel safe experimenting.

Leaders sometimes try to protect people by steering them away from significant mistakes, instead of letting them fail, or at least start to fail. In my experience, if you hold back and see what happens, one of two things happen: either they fail like you thought, and you coach them through it which makes them stronger and bolder in the long run, or they actually pull it off. If that happens, you get an opportunity to learn a new way.

If your employee is headed for a fail, where nobody's going to die, where it'll be a bump in the road and not a fall off a cliff, you let them go.

Re-Recruit (Yes, again!)

People gravitate to the companies with the best working environments, which means those companies can attract the best talent, which results in making the best products. But attracting good people is just the first step.

This takes some humility, but it's important to realize that today's successes aren't guaranteed for tomorrow.

To keep your best people, you need to grow and re-recruit them.

A big part of the environment you create is determined by the people you choose to be in that environment. That means the relationships between you and your employees are vital.

Your one-on-one meetings are dedicated time for your employees, no matter what else is going on. Check in and find out, "What can I help you with? What can you help me with? What can we help each other with?" This is your chance to re-recruit your employees and remind them why you wanted them on your team.

If you want to show your team that they are valued, make time for them.

Your role isn't primarily about the project or the work. Those are specific tasks. As a people leader, your job is to create the environment. You've hired a diverse team of professionals, and now you need to create an environment that each of them needs to thrive and one they can thrive in together as a team. Your job is to set the conditions that let them bring their best skills to the table.

Some leaders may feel they don't have time to do this, that they can barely manage their inbox as it is. One-on-one time is quality time; you need to allocate and prioritize it as such.

That's why this can't wait for a quarterly performance review. Immediate feedback gives employees a chance to start improving now. At the end of an annual review, if any of us is surprised, I haven't done my job as a leader.

But before you can provide critical feedback, you need to build the relationship. The more comfortable employees feel around you, the better they will accept your feedback. You don't get there over email or by phoning in to team meetings. You get that by building an authentic relationship just the same way you'd build any other relationship: face to face.

Meetings matter

So how best do you spend your one-on-one time? Remember that your office isn't a neutral space. It's your territory. Sometimes going for a walk or getting a coffee offers a more relaxed atmosphere.

One-on-ones aren't the only important meetings you have. Team meetings matter too; that's where you built the chemistry and connection among team members, by helping them support each other. But the one-on-ones establish your unique relationship with each one of them. They deserve that if they're actually being asked to bring their unique selves to work.

Good questions are better than answers

When I talk about teaching, I don't mean someone who gets up at the front of a room and just lectures students. I mean someone who brings problems and challenges, and says, "Look, guys. I'm not going to give you the plans of Death Star and say, 'Build it.' I'm going to tell you what problem we're trying to solve with this product. I challenge you to do that, even if you've never done it before. I'm challenging you to grow to seize this opportunity."

As much as possible, I try to focus on asking good questions rather than giving out answers. My employees need me to ask the right questions to guide them towards the right answers. This can feel uncomfortable, because as leaders our employees look to us to have the answers. By asking questions instead, we give up that control. It's not easy.

To get to outcomes, we need to create the right environment, hire the right people, and develop them with the right capabilities. You need your team to be able to think for themselves. To do that, you have to teach them with questions and challenges.

It's like the Socratic method; just keep probing with questions until your employees start connecting the dots for themselves.

This means bringing your team the original problem and letting them struggle with it, no matter how uncomfortable it might feel for everyone (including you). Let them build those skills of asking questions, putting the problem in context, and thinking of a solution. If the problem is beyond your team's capabilities right now, that's helpful too. That will show them what they need to learn next, using that continuous learning budget or project extension you won for them.

Every organization has two products: the one they produce, and the people they use to produce it. I often see that the first product gets more love than the second, but the truth is that if you focus on the people, the product will not only be better, it will constantly evolve. Because you have the right people with the right capabilities: problem-solvers and architects.

I learned that firsthand when I hired the program manager without management experience. I had all these core skills to teach him, and new tools and skills. As he got better, he started teaching me some newer perspectives that had come up since I had done the role.

As an organization, if you invest back into your people, you're rebuilding yourself. If I consume an employee's talent, I get the benefit at the moment and never again, but if I grow it, I get a benefit in the future and going forward.

That's what it takes to be relevant today. Most highly skilled people aren't happy just to have a job. In the global economy there are many new work options and people want to work in the best environment possible. That's why organizations today have to invest in culture, and inclusivity, and retention and engagement.

Exercises

1. Build teaching into your annual performance plan. Tell your employees they're expected to:

   a) train somebody else on one of their unique skills, and

   b) learn the same from another team member.

This way they'll naturally pair up with each other, and you'll make sure that over the years they pair up with different people. As the manager, you might want to help set up the structure, but this exercise requires you to avoid micromanaging. They need to have the space to spontaneously ask each other, "Hey, how'd you do that?" And then spend the next hour learning. You don't even need money for this type of training; just time.

2. Create a list of learning opportunities for yourself as well as your employees; different things that your team members can teach. Include yourself to show that you're open to learn.
    a. Choose something to learn from an employee. Rotate per quarter so everyone gets a chance. This will build relationships with your direct reports and let you practice being humble, curious, and getting others to teach others. Because that's true scaling.

3. Look at your calendar and carve out time for one-on-ones. This should mean giving other things up to make the time. Recall that to say yes to something means saying no to something else.

4. Look at your current and upcoming projects and evaluate if you need to add more time for on-the-job learning and growth. Is your team under the gun from day one, or do they have some room to play with? If they're under the gun, what can you do to get them that breathing room?

Chapter 6.

The need to be decisive

"You can fix a wrong decision quicker than a non-decision" is one of the phrases I always use when coaching leaders.

I've made just as many bad decisions as good ones throughout my life.

And in most cases—sometimes it takes a while—I've come to value those bad decisions, because they helped me make better ones later.

In the corporate world, professionals are afraid of making "career-killing" decisions.

As leaders, it's vital to the growth of your organizations that you recognize and address those types of fears among your decision makers. They need to know that if they are making what they believe is the best choice, using the information available to them at the time, they will not be penalized for taking the risk to make it.

And you should know that, too.

If you have a legitimate and grounded fear that the wrong decision will get you fired, you are working in a restrictive environment. To realize your full potential, you might need to find a new organization to work for—one that doesn't stifle risk-taking.

I can tell you, employees who are rewarded—or at least not penalized—for making the best choice they can with the information they have at the time will be more decisive and feel empowered to take more well-calculated risks.

More importantly, wrong decisions are part of gaining life experience. We don't forget the lessons we learned "the hard way." Those "wrong choices" create huge learning opportunities.

They provide the learning experience-process of trying to get back on track, and allow us the valuable opportunity to reflect later about why we made certain choices, what the results were, and what we could've done differently.

But despite knowing this firsthand, I know it's easy to fall into analysis paralysis when we, as leaders, are faced with monumental decisions.

In this chapter, I'll share some of my best techniques for making solid decisions with confidence.

Making a big decision smaller

The decision to get married is a life-changing one for any couple. Because of my disability, I felt like it was even bigger for Kelly and I. My life was so different from Kelly's, and in ways I felt she could never understand unless she lived it for herself.

I rely on caregivers to help me get through my morning. They come every day. What would it be like for Kelly to have support workers come in and see me naked? To have people in our home every single day. I didn't want her to figure that out after we were married.

We knew we were each other's "One," but the stakes seemed so high. What if it was too much for her? What if she decided she couldn't handle it, and after a year we got divorced?

I wanted her to have a chance to make an informed decision: Could she live with this or not?

So, when we got engaged, we decided to live together for a year before we officially tied the knot.

It was our way of trying to make an enormous decision a little bit smaller. We'd be committed to each other, but if it didn't work, we could part ways relatively easily. Kelly would get all the information she needed to make an informed decision, and she would get to experience this way of life for herself rather than me trying to explain it to her.

Kelly's mom—who is Catholic with strict ideas around marriage—was adamantly against the idea. "Your daughter's going into a unique situation here," I explained to her. "My life is a little different. I want to know if she's okay with it. I don't want to wait until we get married and then realize it's too much."

When we were dating, Kelly went home every night. She was just visiting my life. Marriage would mean that my life became part of her everyday experience. It was also the same for me; I wanted to know if she had any bad habits or quirks as well. And I had so much to learn. She had never even used the bathroom while she was over at my house. Talk about gaining life experience!

"So, these girls have seen you naked?" Kelly asked me when she first met my personal support workers (PSWs)—who happened to be attractive, younger women at the time.

"Uh-huh," I said, just then realizing that might be an issue for her. I hadn't even thought about that.

It made me wonder: What if our positions were reversed, and she had support workers who looked like the Chippendales coming in to shower her, how would I feel?

It was eye-opening. Suddenly, I realized that the decisions I made about my care now affected both of us.

Living together helped both of us learn more about each other. It gave us a real opportunity to discover whether or not we would work together as a married couple. By deciding on the interim step of living together, we relieved some of that pressure and gave ourselves an opportunity to gain valuable information.

Leaders can get better at making decisions

It's important to know that some decisions are a coin flip—they could go either way. We need to make a call and push forward.

We also forget how resilient we are; we lose our inner confidence. You've made mistakes in your life, but you still got to where you are now. You can bounce back from another mistake or two. Every life decision is just an opportunity for growth. Either you make the right decision and move further, or you make a wrong decision and grow because of it. Either way there's no real loss.

There's growth in the process of learning how to make decisions, learning who to talk to, which data points matter, and which ones are extraneous. Deliberate decision-making makes you a better leader.

The cost of stalled decisions

So often, we get caught up in worrying whether we're making the "right" decision, instead of focusing on making the *best* decision with what we know now in order to move forward.

We wait, thinking the right decision will present itself. But you know what? It probably won't.

As leaders, to keep our organizations moving forward, we have to make timely decisions. The key is to get comfortable with the possibility of failure.

The goal doesn't need to be avoiding failure. In fact, any seasoned leader can tell you that's impossible—and not even desirable. The goal is to set up your decision so that if and when you do fail, you can quickly learn and recover from it.

Instead of waiting for the perfect, clear choice, get good at making decisions by seeing them from a new perspective: as learning opportunities. Every decision offers you a chance to keep your momentum, grow and pivot.

The real career killer is waiting to decide, because that's when you lose progress. You've always got to be moving forward in today's world of constant change. You do this by learning exactly how much information and how confident you need to in order to be able to make a choice you can stand behind.

Making the best decision you can at the time is better than a non-decision. A non-decision is when you get into analysis paralysis, thinking "Oh my God, if this decision is wrong the world is going to crash."

It won't. What's more, a wrong decision is valuable if you learn from it.

I've already talked about the value in allowing employees to experiment and also to make mistakes: to fail safely. And as leaders, it's critical that you make space for yourself to fail, too.

It's important to delegate some decisions that are within your team's ability to make. The faster you teach your team that making a wrong decision won't kill them (or their careers), the quicker you all can get comfortable making big and difficult choices.

There is a challenge in allowing the team to choose to go in a direction that you wouldn't have chosen. Too often leaders will agree with the team's decision as long as it's aligned with what they themselves would decide.

Leading for progress requires letting teams closest to the work make the decisions and giving them space to make some wrong ones.

And when the decision does turn out to be wrong, it's important to avoid saying 'I told you so.'

That helps nobody.

What does help? "Yeah, okay. You made a mistake. But you're not fired. We can recover, find the learnings, and quickly apply them."

When my team goes in a different direction than I would've, it expands my learning. It makes me ask myself why I didn't see what they saw. It makes me more appreciative of their work, but it also gives me a chance to ask how I can grow myself.

Sometimes learning is about giving up control over decisions, learning why your team chose an option that you never thought of, and figuring out how to grow yourself to catch up with them.

Of course, even if we learn from them in the end, some wrong decisions can lead to setbacks that are inconvenient, to say the least. And that takes me back to our decision to live together before getting married.

Breaking up work into steps with tighter than usual timelines sparks us to make smaller decisions. Using this structure in business, if a decision doesn't lead to improvement or causes an error, a team can quickly pivot and correct it—even before the rest of the organization is aware that a learning opportunity happened.

That's why communication is critical between departments. It's important for leaders and all team members to be transparent about errors.

It's all part of the overall mindset shift we need to incorporate into our leadership. These learnings are opportunities to amplify an important message: You were wrong, so you pivoted and now you're right.

Celebrate that. It's critical that we stop hiding our mistakes like guilty secrets and start celebrating how we overcame them.

Being open about past mistakes builds confidence throughout the rest of the organization that they can recover from an imperfect decision.

Although I've been talking about right and wrong decisions through this chapter, when I work with organizations, I use an agile framework to move the focus from right and wrong, to good, better and best. There may not be just one good decision. If we condition ourselves to see we're not choosing between one right choice or many wrong ones, but among many good choices to find the most optimal one, then the risk of not making the optimal choice isn't devastating.

The choice you made is still good: it works.

Mid-Chapter exercises

So far in this book, I've had a series of exercises at the end of each chapter. But for this chapter, I want you to do an exercise before we move on. This will give you context for the rest of the chapter.

The first exercise is called the Decision Journal. So often as leaders, we make decisions and then move on. What I ask a lot of leaders to do, to build confidence in their decisions, is to keep a decision journal and review it periodically.

1. Record the decisions you've made in your position, along with your rationale at the time.
2. Write down the results and outcomes. Going forward, I encourage you to make time on a weekly basis to record your decisions.

Once you have your journal filled in, it's time to reflect.

Notice the outcomes: When decisions went poorly, did you lose your job? Did the world keep spinning? As you reflect, ask yourself:

- *Why was this the best decision you thought you could make at the time? What factors did you consider?*
- *What data pointed you towards option A, rather than option B?*
- *What came out of that decision, good and bad? Include anything you learned.*
- *What factors should you have considered, based on any surprises that came up?*
- *Knowing what you know now, would you make the same decision or a different one?*
- *If you had waited longer, would your decision have changed?*
- *Which people helped you make this decision? Did you bring in diverse perspectives?*

When you reflect on your decisions in this way, you should start to see patterns. You might start to see that waiting wouldn't have changed your decision. If it had, your decision might have been very similar.

Just like a food diary or an exercise log, this journal allows you to reflect on where you are today versus where you were. It's to support your memory, because we quickly forget the right decisions we made, but we let the bad decisions haunt us. Often, everybody else forgives us before we forgive ourselves.

When I work with leaders, I remind them that they make decisions every day, big and small, starting with their route to work and what they choose to eat for lunch.

A Decision Journal gives you space to ask yourself, "What have I learned?" Leaders make all kinds of decisions without a second thought. This is a chance to take that second thought.

You have to trust that you've made many of these decisions in the past. You've got the experience. You've got the tools in front of you, including all of the diverse people on your team who can present options you never even would have thought of. Often, it's worthwhile to remind yourself where your data is coming from. Looking at the choices you made that went well or poorly, whose opinions were you listening to for those decisions? Did you take your problem to the team and crowdsource some viable options, or did you keep it to yourself or one or two trusted employees?

When you have a dilemma, it's human nature to go to the people that you feel most safe and comfortable with, the ones who tend to agree with you. You know they'll tell you what you want to hear.

Instead, go to the people who are going to be your opposite. They see the world differently. There's value in choosing a different slate of people to hear from, to ensure you're getting diverse perspectives.

The other benefit of the Decision Journal is that it shows you the impacts and lifespans of your decisions. The good news is that the lifespan of most decisions is shorter than we think. But if you compile all the lifespans of the decisions you make, it adds up fast. That's why the pattern of your decision-making matters. You don't have a time machine; you can't take back the bad decisions you may have made. But you can learn from them and impact the future.

The future gives us a new opportunity to learn from the past. Journaling will let you spot patterns in your decision-making where there's room for improvement. Who would keep making the same bad decisions when the facts are staring them in the face, telling them otherwise?

Is 'yes' a dangerous word?

Yes can be a dangerous word. We say yes to fit in and to look like we're being a good corporate citizen, when sometimes the right answer is no. As leaders we need to be brave enough to say no, if that's what's needed to get the company where it needs to go.

It's important to remember, every time you say yes, you're unconsciously saying no to something else. Maybe it's to your kid's ball game or just being with your family. Maybe it's to another projects.

As a leader, you need to ensure you're properly managing your time.

Think about the last time you got a meeting invite. You looked at your calendar, saw the spot was empty, and clicked 'Accept.' And by filling that time, you declined something else that might be more important when the time comes, but that isn't in the calendar. You removed your flexibility. If you schedule every half-hour block of your day ahead of time, you won't be able to adapt to last-minute situations.

Saying yes is easy. It makes people happy. It's often expected. It can become such a reflex that you don't even think about it. But how can you manage your priorities properly if you say yes to everything?

A better way of making decisions would be to keep a short list of priorities going at all times, and when you're asked to do something, consider how it matches with those priorities or your personal values.

Often, we treat our time like stretchy track pants. We think it'll always stretch no matter how much we fill it. Instead, we've got to treat our time like fitted jeans. Whatever we say yes to, we've got to say no to something else, because we've got to fit in these jeans.

When you go to a buffet hungry in track pants, you're already set up for indigestion later.

If you remind yourself that every yes is also a no to something else, that's a quick way of checking in with your priorities. You might like frequent engagement with your people. How are these low-value meetings you're saying yes to robbing your time from connecting with your team?

It's not easy to say no. In the corporate world, projects come up and often include politics around who is driving them and what kind of relationships those people have with other heavy hitters in the organization. It can seem like a good career move to work with some of these people, rather than thinking about whether the project is really best for the organization right now.

But the truth is, if you focus on what's best for the organization, that's more likely to get you promoted than signing on to a project solely because of who the project owner is.

Some crafty project owners also have a way of making their asks about your identity or abilities as though the way to prove yourself is to take on their project. Then, the first thing that happens after you say yes is that you feel less competent than ever.

I've done this myself. I say yes, knowing I'm going to have to rejig the work plans and wondering who I'm going to have to let down because of my impulse 'yes' to this new project.

This is the kind of thing you'll notice as you review your Decision Journal. Many last-minute urgent projects that come with unreasonably short timelines or large scopes are relatively inconsequential for the organization.

If you look back on the impact of those last-minute asks that you agreed to, you'll likely see it was often minor, at best. In a fast-paced organization, it's easy to forget and move on. A journal can help you analyze and strategize.

You also have to consider your team's capacity. It's not fair to overcommit them because you weren't prepared to say no when it needed to be said.

Studies have found that saying yes <u>undermines leadership.</u>

I keep saying mistakes are ok, but saying yes is an avoidable one. Of all the decisions you make that could go poorly, the impulsive "yes" is the least excusable. To avoid this mistake, you have to learn to give an elegant "no." Let the reason for your "no" provide focus and weight.

Exercise

Add a yes/no component in the Decision Journal. Focus on the things you said yes or no to. Ask yourself:

- *Was that a good idea?*
- *What was the outcome?*
- *What was the impact of that "yes"? What was the "no" that came with it? What was the impact of that "no"?*
- *Was it career-limiting? How long did it put you in the hot seat for?*

One of the benefits of journaling this way is that you realize people tend to move on quickly when you say no. It's usually not a matter of life and death, though it may feel like it. You might notice that when you said no, the project went ahead anyway; the project owner got someone else to help. That can make you feel more comfortable saying no in the future.

Another benefit of this journaling process is that it makes you aware of what's happening in those stressful moments. Before I started journaling, I would roll back to my office and sort of kick myself, thinking, 'What did I just say?' Journaling helped me become conscious of those moments, which helped me stay in control. I got better at catching myself more quickly, and I found by reviewing my journal periodically I got even better at saying no in the moment, even if I came back and was able to say yes later—with buy-in from my team.

A leader should dedicate some work time on a weekly basis to reflect on themselves. The journal is a good tool for this, but you might also spend time thinking about your team. You need to block off time in your calendar to reflect on:

- what you've done,
- what your team has done, and
- how your team is doing.

Importance of time

A benefit of working for myself as a consultant before is that I've had to get better at time management. In corporate culture, you think of your workday as belonging to the company, however they want to use it. Who are you to say no to a meeting?

But if you ever have the chance to work for yourself, you'll see how important it is to use your time thoughtfully. Everything you do should be geared towards improving yourself and business. The same should be true when you work in an organization.

Every request is a request for your time. And it really is *your* time; you need to take it back and plan your day out for yourself, rather than making it available for other people to plan your day for you.

To return to the idea of decision-making: a non-decision wastes time and so does the automatic yes. In both cases, you're squandering progress you could have made. So with every decision you should be thinking, 'Is this the best use of my time?'

The if-we-fail scenario is important. Think it through.

It's easy to think of a decision as a giant monolith with a worst-case scenario.

But often we can actually break that decision down into smaller steps to gather information.

When faced with a huge scary decision, the first thing to ask yourself is how to break it down into smaller stages, and how to use the earlier stages to experiment and gather information.

Have you ever gone cliff-jumping? It's where you jump off a cliff into a lake. It's not really safe, because you often don't know how deep the water is at the bottom of the cliff.

Imagine you're standing at the edge of the cliff, getting ready to jump. It's risky, and it's a huge commitment; once you jump, you can't take it back. You could get hurt or even die.

So how can you test this out first. Can I talk to somebody who's been in that water before? Can I walk into the water, or jump from a smaller height first?

If you have any experience with teaching teenagers, this is the exact type of thinking we're always trying to teach them. *Wait a minute. Think.* It's not that what they want to do is too dangerous *no matter what*; it's that they have to think it through, understand what could go wrong, and then decide how to do it in the smartest way.

Decisiveness is taking the time to reflect on what you know today and what more you *need* to know, as opposed to actually going ahead and doing something in order to learn more. You take as much time as you need to gather the information you can get, and no more.

What helps is to make the smaller decision first; the one that will get you the information you need, or that will move you a step forward. This kind of mini-decision is both a test and a safety net. With the information you get from it, you can take another step forward, and another, until you feel comfortable enough to make the larger decision.

Like trying a sample spoon of ice cream before buying the whole cone. Or moving in together before getting married.

In some organizations, this is called running sprints. If you do tight timed experiments, even a wrong decision only has an impact until you reach your next evaluation period. Once you review your data and build the next version, your mistake will be fixed. It has a lifespan of weeks or months, not years or decades.

When Kelly first moved in, it was a life-changing event for both of us. But because we decided not to get married yet, we didn't feel extreme pressure to stay together at all costs.

We made a less impactful decision while planning to reevaluate in the near future and make a final decision. We knew we were in an experimental phase and could ask each other questions and talk things through without feeling we'd already committed ourselves 100%. We just had to get through a sprint, not a marathon.

We broke the giant decision of marriage down into smaller pieces. Can we even handle living together? Let's figure that out first.

This process is very similar to any goal setting or project management task. The larger project or goal is actually a series of smaller tasks. Every goal or journey starts with something that brings us fear, but when we deconstruct it into smaller pieces, we mitigate that risk and anxiety by focusing on the little steps instead of the big step. You still keep an eye on the big step, but focus on those things that you can control at the moment.

Keep number of goals realistic

Setting yourself up for failure is setting nine goals for the year that all compete for resources and budget.

This happens a lot because of corporate budget cycles. You have to lay out all your major projects at the same time during the budget cycle, to get funding. Once the budget is approved, you want to move forward on as many as possible. It doesn't help that if you have an unused budget later in the year, you risk looking like you don't need it.

But that's not effective goal-setting. Instead, break the work down into chunks. Start the top two or three, and when they're done, then start the next two or three. This way, even if you don't accomplish everything you set out to, you'll have at least a few projects finished and wrapped, rather than having all nine still underway.

Start with two

I'm not saying you can only ever have two goals. What I mean is, you *start* with two. You might have nine, but you're starting with the first two.

This is why I disagree with the corporate budget cycle pressure I mentioned above. I don't like how it creates this pressure where corporations ask, "Okay, what are all the projects we want to do in the year?" Once those are listed and funded, they say, "OK, go and start on all of them." Then everyone scrambles to launch a couple dozen large projects in the beginning of the next budget cycle. Organizations should be making sure everyone pulls together on one project at a time, one after the other. It's better to deliver half the work than half-finish all the work. Whatever's finished can be delivered, but anything that's half-finished is stuck.

So let's say I've got 10 things I'm working on. In reality, I'm only focusing on two or three. Then I move to the next ones and then they become the next #1, #2 and #3 projects. But when we say these are all the top priority, we're pushing the decision-making down to the overloaded front-line workers, for them to decide when they can't fit anything more in. We've got to quit doing that. We've got to give employees a project order to align our priority and vision together.

When you try to multitask, you're refusing to decide what to focus on. Often, it's another convenient yes; you're trying to please multiple people rather than setting an order and working on goals one at a time.

As leaders, we need to start choosing a work order. It can be difficult. It sometimes means telling someone their work isn't your first priority. But it's efficient. Order your task lists from 1 to 10 so your team knows the order. if something else comes up, decide immediately which one is coming off the list.

Take responsibility for those tough decisions for your team, because you're closer to the vision.

If you find the order is wrong, you'll learn for next time what would have been better. And as you progress you can still pivot and reorder the items you haven't started yet. Whereas if you start work on everything, it's hard to reorder it.

Your first decision, when you've got your projects approved for the year, is to order the work. But even after that, there are many smaller decisions around how you're going to do it.

Wrap up exercise

1. Apply decision-making to your calendar, starting with time to work on your Decision Journal. Time for the journal won't fill itself in. Be decisive about your calendar. You get to decide where you spend your time, instead of others deciding for you.

2. Schedule time for reflection into your calendar; about 30 minutes a week. Block off the time like it's a meeting so it doesn't get booked with something else.

When reflecting on your decisions you might find there are no new decisions to record. That gives you more time to check in on the previous decisions you recorded and update them as needed.

This is also a chance to practice and understand your own decision-making framework. It's like any routine; you've got to practice it, reflect on it, find ways to get better. If you don't spend time on it, you're not going to get better at it.

Leaders tend to get busy with everybody else's meetings, and lose sight of personal development goals. But just because we made it to a leader role doesn't mean we stop developing ourselves. We've got to continually get better because the world is getting better around us; the bar is continually rising.

Chapter 7.

Always be in discovery mode

In grade school, I couldn't write my notes like everybody else.

So I learned to type young.

I was in Grade 4 and brand new to Sandwich West Public School, and the blue, industrial-looking Smith Corona typewriter I used in the classroom took up my whole desk. It said "Electric" on the side, but it was noisy, and felt even louder when we were taking tests. You can imagine the looks I got from the other kids.

The poor kid that had to share a desk with me was trying to write between the lines on three-ringed paper, but the typewriter would make the whole desk vibrate—so that caused a little conflict.

It was hard. I wanted so bad to fit in back then. I had been going to a school for kids with disabilities until that year and when I could finally go to "regular" school, I wanted to go to one near my house, where my neighbourhood friends went. But the only school accessible for me was Sandwich West, a more-than-an-hour bus ride from my parents' house in Emeryville.

There were always new obstacles in school. And whenever I encountered one, it would be a race against time to find the solution before the class moved on without me. I had to be willing to try new things—and fast. While the other kids could focus on learning the material being taught, I was also learning a different lesson: how to innovate. I had to in order to access the planned education.

As I mentioned earlier, staying in perpetual observation and discovery mode has been both a necessity and a gift to me. Often, trying new things means watching what others are doing and then figuring out how to adjust and apply those methods.

For me, it's easy to see this as an urgent need.

Corporate leaders should see the ability to change and adapt as urgent, too.

Unfortunately, for many companies, managing change is a low priority.

Continue this thought before moving on or lets work on a better transition

"Discovery" should be the default mode. Instead of being really good at doing the same thing over and over, we should get better at learning new ways and trying new things more frequently.

That could mean changing the way you make your product or service, or how you deliver it. It could mean changing how users interact with your product or even how your employees develop complex work. While learning expenses tend to be minimal, they often feel tough to justify. If you build learning into your fixed costs, it will always be available as part of the work.

Corporations that have a research budget and assign people to challenge the status quo and look for better ways to develop products and services can be better prepared to adapt and adjust.

Implementing a corporate culture that includes a continuous "discovery mode" of better ways to work, helps organizations keep up with consumer expectations in the ever-changing changing world.

So how can companies that are doing well create a desire within their ranks to constantly stay in an innovation phase?

This is the dilemma facing a lot of companies. They find a product or service that's successful, they develop around it, and then they get locked into what they know. The market changes, the customer starts to move on, but the company isn't continuously evolving.

If Blockbuster had been in discovery mode, maybe its leaders would have been ready to compete with Netflix on an evolving playing field instead of trying to push its increasingly outdated video-rental mode.

Why disruption is trending

Apple Inc. is a great example of a company that finds success by disrupting itself. Take the decision to close the iTunes store when more people started getting music through its Apple Music subscription service. Instead of continuing to push the store, which is still successful, or finding money to operate both models, officials started moving ahead to the next trend.

Financial institutions are often guilty of getting stuck in the status quo. They offer about 15 different types of the same credit card. Leaders can be reluctant to kill legacy products, because they don't want the customer to be inconvenienced and forced to move to another one. So, they try to maintain a wide variety of brands instead of focusing on a few and evolving them.

To see how refusing to change can devastate your organization, you need to look no further than the news media industry. Since the age of the Internet, more and more people have been getting their news online, and advertisers have followed. Sticking to traditional news-delivery models, news outlets were unable to compete with changing consumer demands for fast-paced information and thousands of established daily community news organizations in the U.S. and Canada have shut down during the past two decades. Those operating now are still scrambling to keep up and stay relevant now, amid plunging circulation and ad sales. If media organizations had anticipated and adjusted to changing consumer habits earlier, they could have shifted to the platforms where people get their news now.

Being in discovery mode means embracing and pioneering change. It also means realizing that the user/customer is going to change as well. For years, newsrooms counted on the aging demographic that stubbornly continues to demand news in the format it had always consumed it. But as those consumers became senior citizens, the generations who replaced them did not feel loyal to the good old days. They flocked to everything digital, and news organizations should have anticipated and adapted to that been one step ahead so they were ready to provide content on platforms that new potential consumers wanted.

They should have invested in researching their future audiences and ways to keep them. Instead, newsroom managers were smug. They felt their product was what the readers needed and didn't care what readers wanted. They dismissed those platforms as trends and the content as "not real news," and now the industry is struggling to catch up.

You know who could've told newspaper leaders what was going on before it was too late? People running circulation departments, ad sales, and even reporters who have their fingers on the pulse of general public trends.

Again, in any type of organization, to know where your customer's head is at, you need open lines of communication with your front-line staff; the first layer of employees in a traditional hierarchy.

They are the ones who get feedback straight from the customers, and many of them probably are your customers.

Companies often prefer to listen to the business development experts who haven't spent as much time with the users or customers.

When is the last time you brought the whole team in to review users using your product or service? That's how you'll get the purest data: how the customer actually uses the product. If you just listen to what the customers tell you they want, you won't get the whole picture.

Rediscover your employees

Discovery mode also means constantly rediscovering your employees. They're going to change, they're going to make career moves, they're going to grow and evolve. Keep up. Don't keep them pigeon-holed based on what you hired them for. You too should keep evolving, learning, and discovering them. That way, you'll know best how to make use of their capabilities. It might mean helping them grow into new roles as you see a new direction in the market or customer habits.

If you're not grounding yourself as a company, with one foot in your customer base and one foot in your employee base, you'll lose balance.

Yes, I get the irony of me using that analogy, but I think it works.

And feeling uncomfortable around people you might know well is an opportunity to strengthen your communication skills. It forces you to ask things like "How do I get to be right?" and "How do I build their interest?"

Appreciate the imperfect feet of others and how they get back up after a fall

Like a lot of kids, I got bullied in grade school. I was that kid who said "free" instead of "three." I didn't talk cleanly until I was 12. Before that, kids would mock me. But my observation skills were paying off. By watching and listening to other kids and working with a Speech pathologist, it got better and better. I could hear how the other kids were pronouncing everything, I was learning to speak. Things were looking up. Then one day some other kids asked if I wanted to play on the hill with them. And I was like, 'Oh I'd love that.' These kids pushed me up the hill in my wheelchair. I was so happy. Finally.

It wasn't until we were at the top of the hill that I realized the plan. They'd only pushed me up for their own entertainment—so they could shove me down and watch my chair flip over and over.

I remember laying on the bottom with my wheelchair over to the left. I was devastated. I had bruises. But that didn't hurt me as much as seeing and hearing my peers laughing at me. Especially the ones that I thought were my friends. A moment earlier I felt so excited about being invited to be part of the group.

It was awful. But my biggest concern at the moment was 'what am I going to tell my mom and dad?'

I had started to understand that I had to take control of my own decisions. Even though I had bruises from the fall, I remember knowing I could hide it because I would always get bruises standing up and falling. Even though being at that school was really tough, I knew it was a big deal to be integrated into regular school for a kid with cerebral palsy.

It made it worth all the hassle to get there, including the three hours of school bus time to get to the nearest accessible school and the fact that I had to crawl onto the bus to even get to a seat at 10 years old.

Beach

Back on that beach, I started off sitting in my chair, staring at my imperfect feet, wondering how I would ever make my way to the water. Then, instead of continuing to dwell on the imperfections that had always hindered my stance and my walk, I asked somebody to help me up, then started walking.

The sand didn't resist my imperfect feet. It molded around them. The sand changed its shape to absorb my step. It was interactive and responsive. And beautiful.

A good leader is like the sand; being responsive to embrace employee imperfections is an advantage.

Baby steps

I learned to walk later than other kids. When I was finally ready to try, I would watch adults to see how they did it, but they were so practiced and moved so fast I couldn't imitate them. So I watched how babies learned to walk.

Using a couch or table, they would usually start by pulling themselves up, then struggling to keep balance, they would slowly build each skill needed to walk. They fell a lot. Baby steps, right? That taught me the basic principles I needed to start taking steps.

That's when I first learned that to learn new things, we should look to beginners, not experts. Still in the learning phase, the beginner's process is vastly different from someone who's already mastered whatever it is you are trying to learn from scratch.

Think about baking competition shows: If you just tune in at the end, you don't know the steps that went into making that cupcake.

In any organization, we can learn a lot by watching those who have just started something new, and taking note of the steps they are taking to perfect it. Watch the ones in your own industry or another who have just experienced a failure. Find out what happened behind the scenes. Seeing how other industries solve problems can give you a new perspective of looking at that same problem in yours.

Also, by absorbing these stories, you'll bring innovative thinking into your own life. When we look at only ourselves for too long, we start to go blind. It's when we focus on others that we can see things with a new perspective, and then apply that perspective to ourselves.

I love watching new employees or interns take on a new project, when we haven't told them exactly how to do it. I like to see what approach they decide to take and how they do things differently than we usually do. That's how you start to find small ways to innovate.

Suspend judgement

In order to challenge ourselves to do better, we need to always be critical of our own practices. Too often, we fall in love with the way we do things, and feel it's the only way.

If you suspend your judgement as a boss, you can learn new ways to innovate by empathetically observing your employees as they navigate new obstacles.

From watching others, you might find your idea about the "best way" might change. To stay competitive, it's important to constantly test whether the current way is still best by trying other ways and staying in discovery mode. If newspapers had started testing and asking questions about whether their way was best in 2003, the industry might be in a better place today.

As humans, we like to build routines. If we can do something automatically or unconsciously, it frees up our conscious thought for other things. That's a healthy process, but it also means we are constantly reinforcing the status quo, by avoiding the resource-intensive learning process.

I rarely do things unconsciously. My disability means I still have to think about every tiny detail and footstep. That does give me an advantage as a leader. I'm never afraid to try something new, because even the routines I've had forever require conscious effort. They are still new.

But you don't have to have a disability to be open-minded about change. And you don't have to be radical about change to be progressive.

If you encounter a wild new idea or technology, you don't need to jump on it just for the sake of change itself. On the other hand, it's important that leaders don't instinctively, or smugly as in the case of news media managers, dismiss it out of hand. I recommend investigating all new ideas with curiosity. Remember that research budget? It can be used to spend some time doing that investigating to discover whether there is something you could easily apply to your work today.

Bring your electric typewriter

Often we reject possibilities before we've even thought about them. When I brought my typewriter into school, the teachers didn't want it there.

It was too loud. They worried it would disturb the other kids. It was different from how they'd taught the class before. Instead of looking for ways to accommodate my diverse skill set in class, they were trying to resist it so they wouldn't have to change. I had to fight for them to accept that change, prove it would be worth it. It was one of the first of many battles I've had to fight to get decision makers and gatekeepers to agree to what was essentially a pilot project on change.

It turned out that allowing me to bring my typewriter to class didn't disrupt the other students, beyond the slightly vibrating desk. It enabled me to learn and enabled them to learn as well. It also led me to create innovations for others through technology and process innovations, something I continue to do professionally four decades later.

Struggle for change today, to improve your future

The message here is that your struggle for change today will enable substantial returns in diverse innovations in the future.

My stakeholders like to tell me how great things are working in their particular area. People often think it's not theirs, but another area that needs to change. But after acknowledging and congratulating their success, I always suggest they ask themselves what they might be missing out on because of the way they are working.

Netflix does this by taking their whole environment down multiple times a day. They have an initiative where they try to break it to see how well the fail-overs work and how quickly they can recover from a failure. It's an everyday practice, so they are constantly learning about the kinds of crises that could hit their systems. It's a constant challenge and one of the ways that they introduce innovation, that leads to them thinking differently as a company.

Get good at disrupting yourself before a competitor does.

Creating that sense of urgency has to be a deliberate investment. The pace of change isn't going to stay constant; it's going to get faster. Companies need to learn to disrupt themselves and get into discovery mode. The disruptor isn't necessarily going to warn you. Look around and ask yourself this: "If something was going to hit our industry and completely change the way we do things, what might it look like?"

Let the valid fear of being left behind or becoming irrelevant motivate you.

It happens.

Officials at Kodak, for instance, were blinded by their success in the 1970s. They believed film would always be the standard, even though they were the ones to create digital photography. Then they shelved it to protect their film business, and someone else beat them to it.

The idea of trying not to cannibalize your business is an outdated model. Who better to compete against than yourself?

That way, there is no real loser, and you would get the insight to make your next move.

If we don't constantly challenge and try to compete even within ourselves, we become biased by our own comfort.

If you let your bias about the "best way" allow you to rest where you're at, your organizational muscles will stiffen as time goes on. Like my morning bike cardio, you need to disrupt that old way of doing things when it is apparent (or before it is) to everyone else that it is not working anymore for the organization.

As a leader, it's important to keep exercising your leadership skills by constantly changing and evolving to keep pace with your employees. Your employees are growing and evolving because they feel they have to. The team is constantly changing. You should be, too.

Management needs to make sure they put dedicated capacity and dedicated focus on the road map to make this a continuous initiative. If learning is a yearly goal, you won't have to play catch-up. You'll be able to keep pace with change or even get ahead. If you stay in discovery mode, it's easier to make those "quantum leap" business transformation decisions—the ones that will change the face of the company. Your employees will be limber and warmed up, and ready to push together on the new decision. Staying in discovery mode is also good for employee retention. It means you don't have to leave the company to do something new. If you're patient enough, it will come to you.

Expert today, maybe not tomorrow

I am now in a line of work that I never could have predicted, because it didn't exist formally in 1991.

But I got there by constantly learning new skills, looking where there might be a need or an opportunity, and then using them. It can be fun at times, because when you're learning something new, there's no pressure. You get to be a student again; imperfect, maybe even unsure of who you want to be, but open to possibilities.

It's also less stressful to be a leader when you have this mindset. You don't always have to be right, but there is always a quick way to get better. Through the constant pursuit of better, you get right.

Let that allow you to challenge what you think is absolute (when the challenge is only for a point in time).

We shouldn't always be playing catch-up or in crisis mode. Change should be built into your business plan. That's why transformations often don't work. If we're just changing enough to move from A to B, by the time we've moved to B, it's obsolete. So instead of setting a goal post for when you can stop changing, just commit to continuous change.

Even if you try to plan for long-term change, you don't know what conditions you'll encounter down the road. That means you can't really prepare in advance for exactly what it will be, but you can plan for the capacity to adapt to it and then learn from it.

Allow in that childlike excitement at growth and at becoming good at something all over again.

We can't let our fear of failure keep us from trying something new. Allow yourself to be imperfect. Allow yourself that chance to be a beginner. Because the more you do it, the more you enjoy it.

That's why we take on new jobs. Because we get to learn, and the expectation is that we're just starting out in the job. Why can't we let our current employees experience that and get back their enthusiasm for something new? It's like when we pick up a new hobby. We're excited, we just enjoy trying it out... but instead, when we take on something new within the business, we put pressure on ourselves to get it perfect.

It's cliché—but perfect is the enemy of great. The pressure of first-time perfection robs the opportunity for struggle and growth.

If I was a concert pianist, and I picked up the violin for the first time today, I can't expect myself to be in an orchestra as a violinist a week later. But we sometimes expect this from ourselves in business—that we won't spend time in the beginner zone. Eventually we level out the learning curve and get comfortable, and then just do minimal improvements instead of taking on a new instrument altogether.

Being a beginner means not narrowing yourself to only those things you're good at. Instead, you look at what's possible and what you enjoy.

Google is a good example of a company that encourages employees to be beginners. They allowed employees to spend 20 per cent of their time working on whatever research or interests they wanted. They made sure it was an ongoing practice, not something that happened once in a while when people weren't busy.

Some of those projects have been successful, and others they killed off. But the focus was on learning and research. It was ok if something failed.

The problem isn't that companies don't know about the need for innovation. Everyone knows you have to do research and development. The problem is, only development gets funded. We need to give the research aspect more love.

Research is a long-term investment, just like my exercise. I'm not a super athlete today, but as I get older, I'll be healthier and more functional than I would be otherwise. So spending regular time exercising is my commitment for a healthier self in the future. And companies need to do the same thing.

One example of an innovation space in the technology industry is called a "hackathon."

A hackathon is a design challenge that involves putting employees into teams and assigning them something to build by a deadline. The purpose is to stimulate creative thinking by allowing employees to collaborate with different people than usual. What you're really doing is making them into beginners again, with a new team and a new project they've never done before.

To build discovery time into your workflows, the first factor is capacity. If you're trying to achieve growth, it's critical to build capacity for learning into whatever you're working on—whether it's a project, portfolio, or any other initiative that gets budget funding. Alternatively, you could earmark 20 per cent of total capacity for learning and research. Too often, we fill that time with more tasks.

It's important to remember that learning and discovery is not restricted to technology or solutions. It should include all elements that affect your business, including customer behaviours.

Too often when the business has a bad quarter, the learning & development budget tends to be the first thing that's cut. Try to avoid that instinct. Because, in fact, that's when innovation is probably needed more than ever. If you're just trying to do more of the same things quicker and you're not finding ways to do things better or differently, you're not going to get yourself out of that rut. You're just going to hit the wall even faster.

Even if your year-end targets are in jeopardy, try to fight for recognition for your team's learning and development. There has to be an acknowledgement or metric for learning if you ever want to innovate.

Exercises

1. Pick up a new hobby; something you've been wanting to try, but haven't done much of before. The idea is to allow yourself to struggle and be a beginner again. But also, watch how others are learning and learn from them, just as I learned how to walk from babies.
    a. Keep an eye out for other people's imperfectness. This will teach you that it's normal to fall down and get back up. Notice how you react when others mess up as opposed to yourself, because we're always kinder to others than to ourselves. It puts things in perspective to see others struggle; that's the reality, not the chest thumping you sometimes see in corporations.
    b. Once you start to get comfortable, start over again with a new hobby. The more times you allow yourself to become a beginner again, the more forgiving and empathetic you become for others, and that's a

skill you'll need if you're going to foster innovative thinking in your team.

2. Think of a way in your organization to present people with something strange and new, and then just watch how they figure it out. This could be giving your product to a brand new user, without any directions on how to use it. Just observe them. You will likely pick up on details that never occurred to you before.

3. Look for a new routine you want to try in your professional life as a leader. Do you want to try a different tactic or style? Do you want to try delegating decisions? When they bring you a problem, do you want to try asking questions to get them to think what to do instead of telling them what to do? Pick just one and try it out.

Chapter 8.

Be connected to your employee market

High school was a whole new vibe for me. It broadened my mind on diversity and taught me how to build connections and relationships with different people on different levels. I found a group I could belong to, an oddball group of misfits and nerds.

But I was also able to build connections with all sorts of people through common things like video games and music after going back into the observation-mode I had perfected as a child trying to make friends in the playground. Video games? Well, I could kick most people's asses in Donkey Kong. And music helped me connect with the mainstream kids: You know, 'Hey, do you like Depeche Mode? I like Depeche mode!'

I had learned back in elementary school to be curious about others and find commonalities to form bonds. Back then it was stickers. A few years later, it was music. And I also had one other thing: A sense of humour. I learned to have that sense of humour to fit in.

Those relationships are important—even the ones you'd classify as acquaintanceships—because they offer some security and even support to human beings trying to navigate the world. For me, those successful connections were monumental for my self-confidence.

I went from wondering if being disabled would make me excluded to understanding that people accepted me for who I was. Their peace of mind gave me peace of mind. And it taught me the benefit of staying curious and interested in other people, and the foundational skills that have helped me connect with people personally and professionally throughout my life.

Humans are motivated by relationships. They make us feel belonging, which is something we all strive for. That's part of what has made the pandemic challenging.

Good leaders understand that and maintain those authentic connections

It matters. I'm here to tell you that staying connected to your employees should be a key objective for any leader. Of course, being connected means different things, depending on who you ask.

I am the decision maker in the business of my life. But at the same time, I depend on others to help me reach the goals I set out for myself. I have always understood how my connection to others affects my life. Once again, I had no choice.

It started out with family. My parents were my initial caregivers since I was born with cerebral palsy in 1971. Most parents take care of their children as long as they can or need to, and mine took care of me until I moved away for university.

My decision to go to post-secondary out of town marked a huge moment in our relationship. Suddenly, instead of my parents nudging me out into the unknown as they always had, I was venturing out on my own, with their support.

Before moving out, the longest I had ever stayed away from home was for a two-week summer camp I went to every year as a teenager. If it wasn't for my experiences at Camp Woodeden, I may have never ventured outside of Windsor.

The camp had able-bodied counselors who would help me, and it made me realize that I wanted to move out of my comfort zone to go to school. They taught me that there were good people out there who would work with me, and gave me the confidence to move away from home. Some of my counselors are lifelong friends to this day. One of them stood in my wedding.

But even those short trips away gave me a chance to learn who I was without my mom or dad. They were like the sprints I mentioned in Chapter 6: short, safe experiments that helped me build toward a larger goal of living independently. They also helped raise my awareness about what was possible. The camp I went to was for people with disabilities and drew people from all over southern Ontario. That meant I got to see other kids in wheelchairs like me, which I didn't see a lot of in my hometown, Windsor.

When it came time to decide where to go to college, I didn't know what I should major in. I liked both technology and business, and I couldn't decide which to choose. It was 1991 and both had their challenges for me: computer science required lots of fast typing and accounting required writing in a physical ledger.

One day, my dad laid it out for me while I sat at home, trying to decide what to do.

"Let's be honest, you're not going to be a fireman, a police officer, or a construction worker," he started after coming down the stairs and finding me sitting there again. "But you know what else you're not going to be? Living under my roof for free the rest of your life. So you better figure it out."

He was helping me narrow down my options, you could say.

Maybe he truly believed that I could and would live independently. Or maybe he just wanted an empty nest.

Either way, I knew I needed to be employable in a career that would allow me to live independently and to get the kind of training I would require, I'd have to go to school out of town.

But that meant, I needed to access the help I would need to get out of bed, cleanly shaven and showered, and off to class to learn the skills for the job.

First, I moved into a residence for disabled people that offered personal support care.

But after two terms, I moved to a "regular" residence and hired college friends to be my personal support workers to assist. Along the way I had to figure out how to train them, pay them and trust them to take care of me. I gained independence. I learned about management by connecting to others.

While creating an ecosystem around myself, I was providing employment for students and reducing the "administrative" overhead (life management costs) to get me ready each morning.

During this chapter I'll give you some tools to help you understand how to connect with your employees and the benefit in doing so the people who already work for you. To achieve any organizational goal, you need the right people and skills to make it happen.

The first step to connecting to the right people within your organization is knowing what direction you want to go. It doesn't have to be a five-year plan. You just need to have a clear vision of the destination where you want to end up.

Once you know who you are today and where you want to go, you can start thinking about the people you need to get you there.

Like me getting ready to go away to university: once I set a goal to go away for university, I could instantly identify my skills and capabilities gap. I knew I wasn't able to dress and shower myself independently, so I knew exactly who I needed to connect with: people who could help me do those things. Then I was on my way.

Importance of connections

When you're young, the status quo often doesn't seem attractive. Who wants to stay in one place? You're always trying to learn and grow. But as we get older and more comfortable, the pull of the status quo gets stronger. We have more to lose by shaking things up. But it's important to remember, we still have lots to gain.

Every time you meet someone new, you encounter new ideas. And with every new addition to your team, you get new possibilities and new outcomes. Connections are a constant source of innovative thinking, so as leaders we have to be continually searching for different kinds of people, in order to spark new ideas and creativity.

Traditionally, leaders have focused on productivity. But with the pace of change today, we need to move the focus to creativity. And to be creative, you need to expose yourself to new and unusual situations. Think about how your mind grows when you move to a new city, or even a new neighbourhood. Suddenly, you are encountering new situations and people for the first time, with a fresh perspective. You're learning.

Can you think of any encounters that impacted your growth or how you see the world? Did you move from your hometown to attend college? Think for a minute about how you changed and grew as a result of your new environment and the new-to-you people in it.

As leaders, our incentives and metrics are often determined more by what we're doing now, than what we need to be doing in the future. Performance metrics tend to encourage us to do more of the same. When what we really need to be looking at is how you need to do things differently in future to stay relevant.

We already know that whatever's working today might not work tomorrow. So wouldn't it make sense to have metrics that encourage constant growth, new connections, and monitoring the environment around you? A balance between the metrics for today AND future metrics.

You don't grow by doing more of the same. Growth requires constantly putting yourself in new and uncomfortable situations.

More than just skills

We typically hire based on skills. But skill alone won't get you to your organizational goal. As a team leader, you need to recognize that you are relying on people. So consider personalities.

In the business of my life, I have had some support workers that were 100% rock-steady and dependable. I could always count on them to get me out of bed and ready. It wasn't their skills that made them special, it was their reliability and work ethic. When you find the right people, with the passion and personality to engage with the work you need them to do, you are in good shape. You can help them develop the skill.

And your personality matters, too, as you build a team. I was able to find the right people for my personal support workers because they knew what I needed from them and I knew what they needed from me. They were my Team Dave.

How did I know who to hire? How did I get a sense for who would make a good personal support worker? For me, that came down to vulnerability. I would need to trust this person when I was at my most vulnerable: alone and naked and needing to be showered and dressed. Each time I assessed the person I was planning to hire, I kept in mind that in just a day or two, my life could be in their hands.

I had to recruit just like any other employer, and there was no sense in beating around the bush, so my job ads would say things like, "Looking for someone to shower a 20-year-old male. Includes getting him out of bed, washed, and dressed to go out."

It's an accurate job description but it's also required me to be extremely vulnerable up front. That was difficult for me, but it was effective because it helped me get a better sense for the applicants. Because I was honest with them about what the job entailed, I could get their honest reactions in turn, and judge who the best team members would be. That moment of vulnerability, from our first meeting, would allow us to create real trust and a connection.

But connecting isn't a one-time thing. You can't just make a grand gesture of vulnerability, win over your new employee, and then never do it again. As we've talked about before, you must create a relationship that involves honesty and transparency. And you must intentionally re-recruit your current employees by explicitly telling them how they are valued and how they are creating positive outcomes that help fulfill the organization's vision. Remember, too, that inevitably, people will move on and new people will come in and you'll have to start over again, making that vulnerable connection.

And that's a good thing.

Staying connected takes ongoing work

Many leaders will put in some of the work to connect with their employees, then assume that the rest of the work is just going to be done automatically as time goes on. But being connected is about staying curious; not only about the other person, but about where your relationship is.

Do you always talk about work? Do you never talk about work? Do you know not just the basic details of your employees, but also how things are going for them lately? Staying connected is about checking in to see where your relationship is at, rather than just assuming it's where you want it to be. It's easy for a connection to get stagnant if you only connect in one way, or over one topic.

So how does a leader know that their connection has the right breadth and the right depth? It's easy to make it look like you're connected, by chatting with them as you walk by, but when was the last time you were focused on them as a person?

This is something that leaders struggle with. In my work, I often see leaders who are really connected with their peers in management, but they get disconnected from their employees. This isn't healthy. Yes, leaders need vertical and lateral connections, but they also must connect with employees further down in the hierarchy.

Some leaders have told me it's not good to connect too closely with employees. They feel lower-level employees are out of their purview, or that connecting personally might lead to crossed boundaries. Obviously, we never want to encourage crossing boundaries. But as a leader you have to be able to be open and allow your employees to be open. You need to create a shared understanding of who people are, really, not just their work skills.

In fact, studies show bosses who keep open lines of communication with employees are more productive, more collaborative and more loyal to the company.

I encourage leaders to keep an open-door policy and not only encourage feedback but seek it out. On top of that, make space for people to bring their whole selves to work. It's OK to ask how they are doing and how they are coping.

The COVID-19 pandemic has put a spotlight on mental health, especially with many workforces going remote to meet social public health guidelines that discourage non-essential travel and gathering.

But working remotely doesn't have to mean shutting down lines of communication. If anything, leaders need to work harder to actively promote keeping them open. If that means scheduling Zoom or phone calls just to check in, do it. Employees want to feel connected to their work, and often that means being connected to the people they work with.

Because of the way we do knowledge and creative work these days, people have to bring their whole selves to work. You can't maintain those personal and professional compartments. When people bring their whole selves to work, they're bringing everything they've got, which means they need to be able to embrace their imperfections too.

So it's not enough to just tell people to bring their whole selves to work; we also need to be able to receive their strengths and weaknesses without devaluing them. Employees need to feel connected to their teammates and bosses. They can't just feel that they're going to get judged and evaluated when they're trying to be creative.

Without a personal connection to your lower-level employees, you're just somebody who judges them (or their boss) once a year during a yearly evaluation. Instead, you want to be the person who helps them grow, identifies their gaps, stretches their imagination, and gives them new challenges.

Back to vulnerability

To build a connection, you have to be willing to be vulnerable. You have to be able to talk to your employee as a trusted partner and say, "Hey, how are we going to solve this? I may not have all the answers. But I trust you. Let's work together to come up with what we should do and how we should do it to get it to our customers."

Sometimes the greatest ideas come from the quietest people. As leaders, we often default to coming up with all the ideas ourselves, and we create an environment where no one else feels comfortable speaking up. If we really want the best from our team, we have to build genuine connections to unleash those good ideas that are hidden in the vulnerable, less alpha-type personalities.

To forge a bond, you need trust, and some insight into who your employee is—and vice versa. Mutual trust helps you create an environment where people feel safe to take risks and even fail, as I discussed in Chapter 3.

Some might feel exposed connecting with employees this way. It opens you up to judgement, right?

I get it. There is no bigger fear of judgement than the feeling of being naked in front of someone you just met.

That's how I've had to start many days. If I would've let that fear stop me, I wouldn't have gotten out of bed. During the past 25 years, I've been naked in front of a lot of people. Even today, I feel vulnerable about it, but that discomfort fades much quicker than it did two decades ago, because I'm more experienced now.

Authentic connections need a safe environment

When you're genuinely connecting one on one, creating that safe environment, you may be disappointed if ideas don't start flowing from your people. But remember, for many people, learning to speak up and not be afraid of judgement takes time, especially in a team setting. Consider the waiting period an investment—like research and development.

Remember group projects when you were in school? There was always that one person who's more charismatic or confident (or bossier) than the others, and who dominates the group. Sometimes, the rest stop sharing their ideas, go quiet or fall apart into cliques. And as a result, barely anything gets done because no one can agree.

This also can happen in organizations when someone steps up and then forgets to engage the rest of the group. And the culprit could be you. This happens because we leaders may have the power, but we don't always have the answer to the problem. That's why you have to be willing to hire people who don't see things the same way you do.

They'll help fill in your blind spots, if you can convince them you value their opinion. Especially if you can convince them you value opinions that are different from yours. Everyone contributes something important that makes the whole function together.

First, you have to be willing to call it like you see it. Share stories about unpopular or controversial ideas that led to a positive outcome or provided the spark needed to move forward. I can think of many times, when great ideas have surfaced after someone made a suggestion that was way off. Often as the team tried to understand what didn't work about it, we moved in the right direction for coming up with a truly great idea.

Pulling this off is a little easier with small teams. In the past I've run teams that numbered in the hundreds of employees. Today, I run a team that's just 12 deep.

Small teams mean their leaders have a better chance to develop deeper relationships with their members, understanding how to get the best out of them, each individually. But while that is a pro, a con with small groups is you have to more closely manage your team's workload. Less manpower means less output, but not everybody in the corporate pyramid may align with reasonable expectations for you and your team.

Running a small team is like wearing skinny jeans to a buffet dinner. You have to be more careful with what you put on your plate. Being given a large team is like you've been provided track pants, so you can feel free to load up on rolls before you get to the rib eye.

But what happens when you do that? Eventually you're so full you become lethargic. So wherever possible, I suggest organizations develop many small teams instead of a few larger ones because they will be much nimbler, if not outright productive in a gross comparison.

Coaching from the sidelines can help bridge colleague connections

As a leader, when you create these connections with and between your employees, you're always working from the sidelines. You're like my support workers, who do a great job because they are invisibly present.

They allow me to achieve my goals by supporting me from an invisible place. In the same way, leaders who have high-performing teams or organizations are able to help in a way that the spotlight stays on the team. At that point the team is almost its own leadership, and the leader is the one in a support role.

Whenever I talk about serving leadership, I always think of my helper, Bourilack Phommasenh. He goes by B and his dedication makes him one of the greatest leaders I have ever known. When I hired B—who sometimes jokingly calls himself "Worker B"—he was in his early 20s. He was with me through the formative years of my professional career.

B's support enabled me to do work that helped entire organizations change and improve. He was my support worker while I lived in Waterloo, empowering me to do work that made such a difference that it led to me being recruited by a Toronto-based organization within a few years.

Yes, I used my expertise and skill to do game-changing work for my organization. But I couldn't have gotten there without B's assistance, starting with my daily shower and continuing by accompanying me throughout each work day, I wouldn't be able to show up and do the work I do best.

Everybody has that moment in their career trajectory where you really make a huge leap forward, and he was part of that moment for me.

I remain connected to B. When I travel, I hire him to accompany me. He takes the stress out of the trip because of his dedication to supporting me and his attention to detail. We are friends. I even had the privilege of attending his wedding. Now, we work together at Microsoft.

If you recall, earlier in this chapter I mentioned that I spent my first two terms of college living in a residence for people with disabilities. That residence was designated for anyone who needed personal support.

My family thought the supportive environment would help me thrive in university, but instead, I felt like a fish out of water. After a lifetime of living among able-bodied people and being forced to navigate the "regular world," because it was all I knew, it felt like the residence process meant giving up capabilities I had been forced to learn as an outsider with an outside perspective.

There was no more need to evolve and figure out how to do things there, because the residence was made for people with disabilities. Everything was built to be accessible and convenient. But I didn't like it. I was used to struggling in the same environment as the rest of the world. I was used to always having another hurdle to climb over. I had to keep moving forward. I had to keep challenging myself.

Essentially, I had gotten comfortable with being uncomfortable.

But I did have to accept that I have a disability. I couldn't just live alone and accomplish my goals because I wouldn't get out the door. It was because of helpers like B by my side (not in front of me clearing the way), that I could continue being uncomfortable. While B helped with the logistical necessities (showering, physically getting to work), he never went beyond. Aside from helping be my hands and feet when needed, he didn't do things for me. He let me live up to my full potential—and experience the failures that helped me to get there.

B gave me the support I needed to achieve my goals in a world that isn't built for me. My connection with him and others have allowed me to thrive and learn and overcome. Those connections enabled me. They changed everything for me.

To translate this for the business world: the first step to creating innovation is to recognize and accept your current disabilities: your clinging to the status quo, your habit of hiring people who are just like you. You have to help yourself, your team, and your organization get comfortable with their disability, and then escape it.

Genuine connections

Most leaders are great at connecting with employees at the start. They do the recruiting, and after hiring, take time for the honeymoon phase, but then don't regularly reconnect with employees throughout their lifetime with the team or the organization. In today's competitive landscape, with the need for paying attention and building relationships, you need to do it more frequently.

Your role as a leader is no longer about just hiring, breaking down the work, and giving performance reviews. Now, you need to create an environment where you can get the most out of your team so you can get better at managing the team as more than just the sum of its individuals. You have to build your team into a collective where they amplify each other and do more work together than any of them could do apart: where one plus one equals three. To get there, you have to really dig into the discovery phase with your employees, and build a genuine connection.

As the leader, you also have to be aware of how your team members are affected by the work environment.

Leaders are often good at reminding employees about the business context for their work: Why it's important, why it has to be done a certain way. But there is often a gap between the surface—where the employee's business context is—and the deeper dimension of their creative spirit and imagination.

On the surface, employees meet a business need with their work skills. But deep down, they have a deep well of perspective and life experience. To unleash that resource, employees need to feel safe. They need to feel they belong. You need to learn what kind of questions to ask, what kind of environment to create, so they can bring out that part out of themselves. Then over time, you start to step back. The focus is on the collective whole and what they can accomplish, not on you as a leader.

The team handles the work, you handle the team

If you want to grow as a leader, you have to constantly build new connections and put yourself in new vulnerable situations.

You start by forming this ecosystem that I've discussed. Collect your ideal team, get them to work and let their ideas and perspectives start to simmer together. Out of that environment will come results that you may not have envisioned, or even thought possible.

That happens because you have the best ingredients: skilled employees who are empowered and feel safe to learn problem-solving, resilience, and vulnerability. Because the more confident we can be in our imperfections, the more aggressive we can be in achieving goals.

When you've created this environment as a leader, you can focus on what you need to do to succeed. The team handles the work; you handle the team. You get to come out of yourself as a leader, stop micromanaging and sweating the small stuff, and really focus on the big picture. By empowering your team, you're freed to contribute to your highest potential as well.

And that means managing your team like a collective. Yes, you build and maintain individual connections with each separate member of the team, but at the same time, it's important to manage the team itself as a collective. Individuals should not be expected to stickhandle team goals.

As a modern leader, you have to build a whole new knowledge base in assembling your key players, like collecting ingredients for a recipe. You can't just know where to buy garlic. You need to know what ingredients go well together, when to add each one, and how to pull it all together into a cohesive dish with complex flavours that none of them could have contributed alone.

Part of the skill here is knowing what each ingredient adds to the overall dish. You have to be ok with a team that produces imperfect results as part of its process; some ingredients are tricky to combine and others will be bitter at first. It takes time for everything to simmer all together to build the final flavours. If someone comes into your kitchen to check on you and tries your dish before you've added salt, they're going to think you're a terrible cook.

So you have to be confident in the uncomfortable interim stages before the final results are ready—where things are messy and incomplete. You have to be really tuned in while your team is combining their work and ideas, to know how things are coming along and what new ingredients need to be added when.

An example of this is when my career advanced to the point that I started traveling. Up until then, I had never thought I would need a personal support worker for trips, because to do my job, I just needed to get out my door and into the office building.

Then my career progressed to having global responsibilities, which meant I had to learn how to travel for work.

And that meant new conversations with my employer to ask for travel reimbursement for my helper.

Beyond those conversations, I was nervous about the trip itself. Work traveling is extra stressful, because instead of having my wife and friends as backup, I have to fully depend on my helper. My life is in this person's hands. I had to embrace that fear if I ever wanted to see what I could achieve on a business trip.

Connections are vital: They spur growth, creativity

Building new connections and encountering new challenges, should never stop as long as you're growing. After going away for university, I got comfortable with having support workers help me. After moving in with Kelly, I adjusted to living with her and having her be part of the equation. Over time those things got easier. But a business trip meant I had to adjust yet again, to being alone with a helper in a foreign country. Each step of the way, I had to give up the comfort of the step before. And each time, I did it because the new step was worth the discomfort.

There was a time when I was afraid of vulnerability, like many of the disabled people I know, and anyone else, really. But I've seen how being vulnerable with others has opened doors for me that would've seemed locked to medical professionals who first recognized my disability.

Once I embraced what felt like it would tear me down and pushed beyond the status quo, I found being open about my vulnerability actually made me stronger than I ever thought I could be.

The status quo is always chasing us. It's up to us to just try to continue to keep ahead of it.

The year I moved away for university, I found out I'd been accepted months before it was time to go. By spring, I knew roughly what was going to be happening in the fall; where I was going to go and when I needed to be there.

But there was no sense in planning out most of the details months in advance. If I had let myself get into all of the logistics and unknowns and potential roadblocks too early, I might have talked myself out of it.

So that spring, I took myself through the more simple questions. What would it mean to be away from home, without my mom and dad to help? How would that impact my day?

Right away, I came up with more questions. How will I get out of bed? How will I brush my teeth? How will I shower? How will I dress and get to class? My old support system was my parents. I won't have them anymore. How do I replicate that system? Who are the kinds of people who could help me with these tasks? How much would it cost? Is there funding available?

Of course, each question was a potential rabbit hole of unnecessary details and fears. When it comes to hiring someone, what if I hire a murderer? Or someone who has a weird fetish for disabled people?

But I knew better than to get distracted by those fear-based questions, so I didn't let my mind wander. Instead, I stuck to surface-level thinking and focused on the big picture. Avoiding rabbit holes kept me flexible and curious about the larger possibilities, so I could brainstorm ideas for my future life away from home.

You're not always ready for the needs of the moment. I've rarely been ready for any new step. It's okay to experience fear of failure. Try to let that motivate you rather than cause you to shy away.

If you can't push through your fear, the status quo will catch you. It can be a comfortable prison. That's why when I reach a certain level of comfort, I push myself to try something new.

Broke student mentality

You may think you don't have to be resourceful anymore—that you have everything you need. That's the lull of the status quo. Put yourself and your team in positions where you're occasionally challenged enough that you have to be resourceful. See what happens when you have nothing; this will show you how strong the human ability to make something out of nothing is. Challenges like this will also show you the connections you need to build, since you can't just do it on your own.

Sometimes software companies do "stress tests," which determine how robust software is by pushing it beyond it beyond the limits of its normal operation. Organizations can do those types of tests in other areas as well. If you give your team a "stress test" challenge, you will find areas where people or the team as a whole needs more resources or training and areas where the team is skilled at coming together to solve problems.

Think about how students do projects vs how companies do them. When you put a group of students together to build a project, they are resourceful and do it in the most affordable way. They will often find resources that are free or close to it that will enable them to collaborate together on—and create—their project. Because they are hungry to get it done. Or to prove they can. Organizations lose that hunger.

When large companies come up with a new project, they often price out the most expensive way to do it, and then say they simply don't have the budget.

Adopting a "broke student" mentality makes it cheap to try new things, and cheap to fail. When you make failure affordable, the opportunity to grow and innovate significantly expands.

Running a stress test or challenge or hackathon with your team every so often solves a few different problems. It helps your team become agile. It allows them to put creativity first instead of productivity. It forces them to make new connections. It strengthens the connections they have right now. Because whether you succeed or fail, you're working as a team, and you get to see the resourcefulness of every single person on the team.

This type of exercise challenges both your team and your individual employees on their creative courage, their resourcefulness, and connections. That's how you start to create an environment where one plus one can equal three.

It's kind of like traveling to a new country. Everything is different—the language, the culture, the people—and it takes away a lot of your normal habits, which helps you rediscover who you really are. That sense of discomfort fosters creativity and the chance to reconnect with yourself and others at a different level.

Pre-exercise Amble:

Before I get into the exercises, I want to note this chapter has emphasized focusing on the internal market: your employees. The external market is important, but that's typically where the focus begins and stops. We're trying to focus on the people who are normally underserved, the employees.

Without employees, without people, we don't have the means to do anything at all. Corporations are a collection of people. And a corporation is only as good as its culture. If your organization doesn't have a good culture, it doesn't have a good heart. That's why employees must come first.

Exercise #1

Find a way to make yourself vulnerable to your direct supervisors. Expose something that you wouldn't normally want them to know. Obviously not something that would get you fired. Consider sharing a characteristic or personality trait that you know holds you back at times, something you might even regard as a disability (within the context of this book).

If you happen to have a real disability or difference, you could discuss that if you want to. If not, look into your undiagnosed disability and diagnose it. Part of the benefit of sharing this is that you might also gain some insight into the ways your "disability" is really a strength. I know my disability has made me who I am, and it's given me the talents that I have. It's been one of my greatest inspirations, and I didn't see that until I allowed myself to be fully vulnerable and accept it.

Another kind of disability could be a trauma or past experience that holds you back today.

Does your leader have to reciprocate with their own disability? Not necessarily. After all, I don't encourage my helpers to get naked in the shower with me. I mean, that would just make it weird.

Exercise #2

Now do exercise #1, but with your employees—share something that will make you vulnerable. This is one of the ways that you show them it's okay to be vulnerable—that you're right there with them.

Exercise #3

Take inventory of your relationships with your employees and ask yourself, "Do I feel connected?" As you're considering your relationship with each employee, notice the emotions that come up. If you're feeling a lot of resistance, that can be a gauge for how much you're disengaged from that employee.

Exercise #4

Take some time to reflect on the effort you put into understanding your customers and users versus understanding your employees.

How much time have you spent getting to know your team? Have you spent an equal amount on your team as compared to your customers? Have you been putting as much focus on your business as you are on your employees?

More often than not, the answer is no. That's likely because the traditional work culture discourages us from getting close to employees. It treats the employees as a means to an end, so we're taught to keep boundaries between us. Obviously we need boundaries, but they need to shift to allow people to bring their whole selves to work.

Your employees have many, many dimensions that you don't know about, and these represent opportunities for you and your organization. They have all of these untapped abilities in their personal lives that you may not learn about if you're only thinking of them in terms of what they were hired for.

I only need help in the mornings, so my personal support workers often have second jobs or school in the afternoons. So in my reference letters, I always say something like, "Just to give you an idea of how responsible this candidate is, if they did not come in at the right time in the morning to get me ready, it would impact my whole day and the day of everybody I interact with in my role as vice president.

" If they can have the responsibility to hold somebody's life in their hands, and know that that person's day doesn't start until they show up to work, then how much better will they do for you when their responsibilities really aren't so mission critical?"

So my question for you as a leader is this. How well do you know your employees? Would you know if you have a superhero in your midst like one of my support workers? Do you know who else they give their time to, and what those interests or side gigs say about them as a person? Isn't that at least as important as making sure they're making the widgets you asked for? This is the kind of thing that should go in that team management Lean Canvas that I mentioned earlier.

Group exercise

Have your team members list "invisible" skills that the rest of the team may not know about because they are not related to the job.

There's a difference between a role and potential and capability. Staying connected allows you to see all those layers.

Exercise #5

Ask yourself, what is one project or initiative you've been wanting to do, but couldn't get the budget for? Turn it into a tabletop exercise for your team. Get them in a room, lay out the problem, and tell them they have to solve it like they're students working on a group project with no budget.

See what they come up with!

In solving it—or attempting to—new characteristics of your team collectively and individually its members will emerge. It's a chance for everyone, both you and team members to see themselves with new eyes.

Chapter 9.

Get great at being imperfect

In the house where I grew up, in order to get into the kitchen from the family room, you had to go up five stairs.

For every meal, I had to climb up those stairs on my hands and knees.

It was a struggle, and I felt like it was something that should be fixed in a house that had someone with cerebral palsy living in it. Many times over the years, I asked my dad to build me a ramp to get into the kitchen.

He never said no. It was always 'Yeah, yeah, maybe this weekend.' or 'I'll get around to it,' or whatever. But he never built the ramp.

I went through elementary school, high school, graduated and moved away for university—all with those kitchen steps in place. Then, after graduating university, I moved back home because I was short on funds, and it was still the same. Not exactly what you'd call "accessible."

'I'm not always going to be there to build ramps for you'

One day, dad and I were sitting together at home busting each other's chops and I reminded him of the time he told me I needed to figure out what I wanted to do with my life because I wouldn't be living under his roof.

"Aren't you glad that I got you to follow through and do something?" he replied.

I really thought I had him, then.

"Yeah!" I said. "About following through... All through my childhood, you always said you were going to build a kitchen ramp for me and you never did."

He didn't miss a beat.

"Well, I did that on purpose," he replied. "I'm not always going to be here to build ramps, so you're going to have to learn to struggle on your own."

Two days later he died.

When we had that conversation, I was a young man and intent on masking my struggles. I didn't want people to know how difficult things had been and still were for me. I wanted to pretend my life was smooth and I had it figured out, when really it was a constant battle.

Crawling into the kitchen on my hands and knees was such painful, and glaringly obvious, evidence of my limitations. I didn't want to face it.

It took a lot of time and energy to put on that facade of keeping up a perfect appearance. I wanted people to look at me and think, "Look at this person. Despite having to use a wheelchair, here he is," when the truth was, I was still trying to figure it all out.

I still am. I don't lead change in organizations and give talks because I know everything. I lead change in organizations and give talks because I know how to struggle.

But for a long time, I held onto the idea that I could hide my imperfections. In my early career, I feared revealing my struggle would hold me back professionally. When I met my wife, I was afraid that if I let her see how hard it is to live with a disability, she wouldn't want to be my partner in life. I was living my entire life the way I presented myself on a first date: on my best behavior, masking my flaws, trying to uphold this image of how I wanted others to see me.

Typically, we stop doing that after a few dates, because it's exhausting and not how anyone wants to live.

It took me longer to drop my guard.

I didn't grasp the full meaning of my dad's words until years later, when I started actually embracing my challenges and imperfections, and allowing myself to continually struggle up and down whatever stair climbs lay in my path.

Life: a work in progress

My dad died right after I finished university. One of the things I wish most is that he could have seen me succeed.

He and Mom had worked so hard to get me through school. It was a constant struggle.

They fought battles to give me a life they could imagine for me, even though they knew it would never be perfect—without knowing how I'd get there. Whenever we solved one battle with the school system or whatever, there was always another one, and usually extra expenses. My dad owned a heating and cooling company and I remember him working on rooftops in all weather—from winter frost to Windsor's scorching summer heat—to make money for my education.

Dad never got to see what I built with the foundation he gave me. He didn't get to see me do well in my career, fall in love, get married—have a full life. He saw the struggling along the way, but he didn't live long enough to see me really hit my stride. But it paid off. Twenty-five years later, what he instilled in me has allowed me to help others to struggle and climb up their own stairs. I'm here and life keeps getting better.

Growing up the way I did taught me that life is always a work-in-progress. No matter how dedicated or resourceful you are, there will always be another challenge.

Modern leadership

When I think of real modern leadership, I think of my parents' example. They started with a vision for what my life could be, even though they said they didn't personally know of any nearby role models of "successful" people with disabilities at the time. And even though the experts—doctors and teachers and school principals all told them I couldn't live a normal life.

They went with their heart. Sometimes you have to do that. And sometimes you just have to go with your gut.

When the world says NO, true leaders ask 'Why not?" When imperfection is the challenge, they embrace it. They get comfortable with the fact that they don't know exactly how they'll make every step to their destination. They know if they start walking and don't stop, they'll get there eventually. All my parents had was a conviction that I had the potential to live a full life. Each fight aimed to win over people who thought their vision was impossible and were refusing to help.

It took me a long time to learn to lead like my parents. I spent much of my early career trying to mask my struggles, past and present. Once I finally learned to embrace my vulnerability, I could inspire others to persevere.

No leader has everything figured out. The minute you think your struggle is over and you've "arrived," that's when complacency and the comfort of the status quo take hold.

This book includes metaphor and theoretical examples because while living an imperfect life challenge-to-challenge is a concrete reality for me, it isn't for everyone.

So my challenge to you is to find a way to apply the ideas and concepts in this book to your practical reality.

I *know* your organization has some entrenched thinking. I *know* you're in a bit of a rut. I know that because it's true of everyone and all organizations. What I don't know is what that looks like to you, or what steps you need to take to embrace imperfection.

I also know many of us arrange our lives to avoid the things we're not good at or can't do. And we often think doing something imperfectly means we're not good enough.

Here's a tip: If by some chance you have built yourself a life where you aren't living from one struggle to the next, it may not be because you've got it all figured out. It might be because you're not dreaming big enough. Most people are not naturally great at being imperfect. It takes a long time to get comfortable with living as a work in progress.

Let me ask you to consider these questions:

- What would it be like if you embraced a growth mindset? Not just at work. All the time.
- What if you started taking on challenges and opportunities everywhere, much like a child does, instead of avoiding situations where you might fail or lose face?
- What areas would you want to grow in? How would you want to do it?

Try to catch yourself in those moments when you're worrying more about your image than your growth as a person. You can devote your effort to keeping up that perfect image or you can reallocate your time, effort and resources. Use your imperfections to keep your edge, to keep your hunger, to keep your eyes open wide.

Just take that step

As a kid, I was really invested in the idea that crawling up the stairs into the kitchen on my hands and knees was not what success looked like. I wanted a ramp so I could stay in my wheelchair.

When my dad said he wouldn't always be there to build ramps for me, his point was that I shouldn't shy away from the struggle. I needed to work through that struggle and learn to build my own ramps or get by without them—or buy my own house that didn't have stairs. I didn't need an image of success that someone else built for me. And I certainly didn't need to shy away from the feeling of awkwardness that I got from crawling up the stairs. As it turns out, that's what struggling and embracing your imperfections feels like.

Success isn't a permanent state. It's a snapshot of time. You accomplish something and for a split second, you're done. But you can't hold onto it. Almost immediately, the next opportunity or challenge presents itself. The only way to build ongoing success is by continually finding the next imperfection to embrace, the next frontier of discomfort to enter into.

If you think about what it really looks like to walk, you may be moving in a straight line but each step is actually a tiny departure from that line. A little to the left, a little to the right. Then you adjust back to centre. The moments where you are actually perfectly in line and in balance are just a tiny fraction of a second.

Imperfection isn't something you can overcome or eliminate, once and for all. Today's perfection is tomorrow's imperfection.

I'll never overcome cerebral palsy, but I'll learn to live with it and to make the best of what it is.

If you feel you've truly reached perfection, then you're missing two important things. First, in order to be perfect now, you must feel there's no opportunity to grow. How sad is that? Second and more importantly, in order to be perfect now, you must have eliminated imperfections from your life or your organization. But imperfection is reality.

Imperfection is something to be celebrated. If you insist on perfection from those around you, you hold them to a very narrow standard for performance. If you allow your team members and employees to be imperfect, it gives them infinite options for how to do their best in that situation. And their best, even if it looks like failure at first, is always better than not trying at all.

Don't just accept your imperfections, embrace them

It's important to draw a distinction between acceptance and embracing. Acceptance is passive. When people say they accept the fact they are imperfect, it sounds like they are fine with the status quo. When you embrace something, it's active. It leads to you take new actions. That's what should happen, because most of us grew up with the idea that if we can't actually be perfect, we should at least be trying. We don't talk about our struggles at work—not until they're far enough behind us that it's comfortable. But being imperfect is what we're all naturally good at; it's our true state. So if you are embracing your imperfection, one of the first things that will happen is you'll start to make life changes and changes at work.

If my parents could have created a perfect life for me, I wouldn't have had to struggle. I would have been able to accomplish my successes easily, and all of their effort and resources could have been spent improving their lives instead of mine. But my struggles, setbacks and missteps have been the journey to where I am now—and given me a story to help others see that it is possible to change and adapt.

Many times in meetings, I've heard someone say, "We want to be the best." But being the best, or even great is just a moment in time. There is always a new great and a new best on the way. If you only look at grandiose ideas of what will make you the best or the greatest, you'll overlook all the immediate, concrete opportunities to get better.

When my dad told me he purposely chose not to build me a ramp to get into the kitchen, it seemed a little dismissive, like a convenient excuse. But looking back, I see how that purposeful choice on his part prepared me for the lessons I would learn later, through situations like my first job and that walk on the beach.

I wanted the ramp so I could avoid using my imperfect hands and feet. But my father chose to keep me grounded in the reality of experiencing my body as it really was. By preserving those moments for me, instead of taking them away like I wanted, he helped me get to a place where I could lean into that struggle.

The staircase never ends

In that story, I only had to climb five stairs. But the bigger picture is that if you embrace your imperfections, the staircase never ends. The first time you look at it, it can seem impossible. It only becomes possible as you climb more and more. And it only becomes truly impossible once you give up. As you climb, you encounter new issues. What worked before doesn't work now. Your knees start to hurt. Your hands start to hurt. So the climb is continually changing as you find ways to keep going. Over time, your muscles get stronger. Your limits increase. You gain confidence.

That's what happens when you embrace your imperfections instead of rejecting them as if they're not a part of you. When you're caught up in how you're perceived, you really want to distance yourself from any imperfection. It's not you. It was just a one-time thing. There's an explanation. You don't have a problem.

And that's how perfection, or the pursuit of it, weighs you down and keeps you from actually trying. As I've said, perfection isn't a state you can actually achieve, because it only exists for a split second. What it really is, is an excuse not to try. If you *say* you made it, you're perfect now, then no one can expect you to take on a new challenge.

But all that leads to is you hiding your imperfections to keep up this pretense. And that's when imperfection becomes a powerful secret that can be used against you. But if you embrace it, and get comfortable with it, nobody can use it against you. If anyone tries, you just acknowledge it: "You're right, I am imperfect. Let's try to make it a little better. So instead of you judging me, let me make you a part of what I'm working on, and let's figure it out and achieve it together."

Both perfection and failure are temporary states. The trick is to bounce between the two, not staying in one state too long. When you're in a failure state, learn from it and try again. When you're in a perfect state, learn from it and seek new ways to fail.

Build the conditions where it is cheap to try and quick to respond

Perfection is a dangerous metric. If you really believe you can make a perfect software solution, you'll invest too much time and effort to achieve that goal. You'll risk everything. That's how perfection holds you back, blinds you to your true reality, and fools you into chasing it.

Perfection is seductive. When you feel you've done something perfectly, you get an endorphin rush. That gives you a temporary sense of confidence. But the only thing that will help you build real, sustainable confidence is failing and learning.

In an agile culture, we do things in small increments, and that makes failure affordable. And once you can afford to fail, then you get the opportunity to invest in growth.

The reverse is also true. Until you can afford to fail, you'll never be able to invest, grow and learn. Failure is the barrier to entry for continuous learning and growth.

Everybody wants to succeed. But if you're not prepared to invest in potential failures, how can you succeed? And if failure isn't affordable, where you can fail and still recover, then nobody can afford to get into the game. We've got to lower the price of entry by making failures smaller. By embracing failure, we're investing in that opportunity to get better. Not to be perfect, but to continually be better.

Don't fear failure so much!

Occasionally, true failure comes along. It hurts and it's damaging. How are you supposed to recover from that?

My advice: let yourself hurt. Feel it. You can't deny that it happened. It's going to take time to recover. So the goal is to actually allow yourself to sit with that emotional state and learn whatever you need to learn, but then say, "Now what?"

The key word is "and." Fail *and* do something about it. Failure and success are not mutually exclusive—not either/or. Think of it this way: I failed and also I will succeed.

Making failure affordable is one of the best ways to reduce your fear of failure. Look for ways to fail where it won't be as impactful. Force a quick turnaround on your next learning.

It's tempting, with a high-impact project, to just delay checking in and getting status updates or to put blinders on, say everything's fine, and push ahead. If there's only a short timeframe and then you need to collect feedback and make the next iteration, then there's no time to hide, and no time for things to get way off track. You'll quickly notice what isn't working, before it becomes a disaster, and fix it for the next version.

Smaller projects with tight deadlines allow you to fail often and get used to it. And again, because it's affordable you can.

Designing a project with quick, small failures is also a great way to train yourself to live with imperfection. It becomes your new reality. Once you get to the point where you don't let failure or imperfection get in your way, then nothing will. The quicker you make smaller imperfections, the quicker you can adapt from it towards perfection. And if your organization can embrace imperfections, they will learn that failure is just a means of creating success, and they'll be ready to innovate and pivot as a modern organization.

Failing fast doesn't mean you never take big risks. That's not my point. My point is, take calculated risks. And be prepared to acknowledge the failure fast, because the quicker you do that, the quicker you can start the learning process.

Many leaders, consciously or unconsciously, try to blind themselves to their behaviors or habits that might be creating stress and what appears to be an unsustainable work environment for their coworkers or employees.

It's the kind of mistake you don't catch until it's too late and you've already lost employees. But if you're going from one small iteration to the next, your role as a leader is clearly defined. You have to be able to accept what isn't working could be your leadership. By failing fast, you can find those errors and fix them before your employees start walking out the door—talk about a failure you can't recover from.

Embracing failure as a mentality will help you build confidence. If I have one success, the confidence I get from that isn't sustainable. It's only sustainable as long as that success is relevant. But learning to fail, continuously embracing imperfections, helps build the sustainable confidence you need to deal with adversity.

Be the chef

One of the things that you notice about a really good chef is that they're often completely comfortable with failure. You can see this play out on competitive cooking shows.

One contestant will notice that their dish isn't shaping up the way they want, and they'll quickly change their plans. They might decide to throw it away and start over. They might repurpose what they've got to make a different dish than what they originally intended. They'll improvise. And the chefs who can be really fast and objective with these decisions often do well in the competition. The chefs who stick with their failing dish, trying to get it to work, are the ones who run out of time and get eliminated.

An innovative environment needs an experimentation mentality to thrive. There needs to be a willingness by leaders and team members to notice what's not working and fix it right away.

Imagine a team that worked like this—members looking out to discover what's not going well, bringing it up, and figuring out what to do next. No time wasted on trying to keep up the appearance of success or uphold the status quo. In an environment like that, most failures wouldn't even feel like failures. They wouldn't sting because they wouldn't need a lot of time or resources to fix. A team that worked this way would quickly become braver and bolder.

Back to the future

In fact, we all started off this way. Do you remember being in grade school and starting a new subject? You'd start off knowing absolutely nothing about it, and then over the course of the year, you'd do assignments and take quizzes and study for tests, over and over again, and by the end of the year you might really love that new subject and feel very confident about it.

High school was always a struggle for me. My grades never came easy, I had to work hard for them. I had friends that got great grades with putting minimal effort in. And I remember the guidance counselor telling me;

"You watch. In university these people are going to struggle."

'Yeah, whatever,' I thought.

And then sure enough, in university my grades stayed the same, while many peers who had gotten good grades in high school were suddenly struggling. In high school, they'd focused on subjects where they excelled. But in university, they needed to get credits in different disciplines and many of my peers didn't have the tools to persevere.

We were all taught to pursue good grades in high school. So, it made perfect sense to stick to the subjects you were already good at. If you chose something that would be challenging, even if there was a good opportunity for personal growth, your grades would probably slip.

It wasn't worth it. That's why so many kids graduate today without those crucial coping skills. They don't know how to fail. They don't know that failing can be good, because you can learn from it; that's not what our education system tells them. Not when it comes to grades.

When people learn about themselves and how they handle themselves in the worst conditions, it better prepares them to be able to move toward the best conditions. It makes them braver. If you've been through hell you aren't scared of losing the appearance of perfection.

When you expose yourself to failure, you learn that whatever is happening now isn't forever. You learn to let go of the past and look to the future, planning how to take your next steps up the staircase.

Failure teaches you to be confident in yourself. Once you've been through a few major failures, you know that you'll come through it.

You can get back up quickly, and you can empathize with the others around you who are getting back up from their failures. You learn how to recognize those special talents and resources in your organization who are good at failure in this sense; who can go straight to asking, "What now?"

As leaders, we have to model this comfort with failure for our employees, and help them learn what might be a new skill for them. We have to help them recover quickly, to see that it can be done. We have to make them feel safe to fail, and then stand by them as they struggle their way out.

It's not for us to build ramps for them. We should be motivating them to start climbing back up the stairs.

But what if the failure is a human mistake, not a risk?

So far, we've been talking about experimental failures. You tried a new thing and it didn't work. That's relatively painless. But sometimes, the failure comes down to human error.

Kelly works in payroll. Recently, she started with a new company, and while previewing something, she accidentally sent the payroll off to the paying company. It was human error.

But it was also a design error. There should be more than one button that you hit to send payroll for payment. It shows that even human errors can be an opportunity to grow and learn and error-proof that faulty system.

The way we design processes is really important. Especially with a repetitive process like payroll, it can become so mundane that people start to do things without thinking about each step.

That's good; they're becoming efficient at using the system. But you need red flags in place to get their attention and help prevent costly mistakes.

If you want to prevent failures, start by looking at the system, not the person. It's easy to blame an employee for not paying attention. But a failure of human error should still raise a question of, "What now?" Now that you've learned the system is subject to human error, how do you fix the system?

In a healthy organization, new systems are coming in and old ones are constantly being re-evaluated. That's why you should expect to see failures if your organization is changing and growing. If you don't fix repetitive failure, you don't get the time or capacity to do new failures.

One important way to reduce your fear of failure is by inventorying your strengths. That way, you know what resources you can draw on. To return to my story of crawling up the stairs, after a while I developed strong muscles from doing that every day. As I grew bigger, I had to learn to adjust. Movements that worked when I had short legs didn't work when my legs got longer.

But really, I was learning the mechanics of climbing. I was building a skill I could draw on for the next time I had to climb an unfamiliar set of stairs.

If my dad had built me a ramp when I asked, I never would have learned to climb. If you shy away from failing, you never get the benefit of seeing those skills and talents that get you through it. You never build the sustainable confidence that comes from knowing what you can handle. You get a sense for the resources that are available to you.

Gold powder in the glue

When you fall, you learn how to put yourself back together. Often, we want to put ourselves together so we don't look like we ever fell; we hide all the damage. But that's not the only way to do it.

There's a beautiful practice in Japan called *kintsugi* where when you glue a broken cup or bowl back together, you use gold powder in the glue. When the piece is repaired, the cracks all show, but they're beautiful. They become part of the new piece, and it becomes a work of art.

I believe this is a perfect metaphor for embracing your imperfections. Don't try to hide your scars. Celebrate them.

Fail and . . .

When we fail, there's two mistakes we make.

Sometimes we try to justify the failure too easily or make an excuse for it. We want to let ourselves off the hook. But when you do that, you can't learn from it. It's important to let yourself feel like shit for a while. You've got to get used to going through that painful thing. If you prevent yourself from feeling the full pain of failure, then you don't develop empathy for others, or grow.

That's why it's so important to fail <u>*and*</u>. Fail *and* learn, fail *and* grow, fail *and* try something else. That "and" is your life preserver to get through the pain and keep going.

One of my favorite movies is *The Crow*. I loved it when the main character Brandon Lee says "It can't rain all the time."

Weather, bad or good, is temporary. You can't truly enjoy good weather without going through the bad. The same goes for successes. You can't truly enjoy them if you haven't experienced failures along the way. Embrace the imperfection.

The present moment can be the worst we've ever felt, or the best, depending on what we've just experienced.

You must keep climbing.

Part of the struggle I had with the kitchen stairs is that I really wanted to outgrow that ordeal. It was something I'd been doing since I was a little kid. As a grown young man, with nice clothes, I didn't want to get down on my hands and knees. It didn't fit with how I saw myself. I'd changed.

My overall goal was still the same: I wanted a full, independent life. Crawling up and down stairs was good enough when I was a child; it allowed me to join everyone at the table. But now it didn't feel good enough. What I needed was my own place—with no stairs, or with tools to manage the stairs. The goal stayed the same, but I had changed, so I had to adapt my approach. Achieving that goal meant something different now that I was an adult.

If you are working on embracing imperfection, you'll start to notice others in the organization who are doing it, too. They'll impress you because they're not settling for a pretense of perfection; they're really embracing their imperfections. And the beauty of being imperfect is that you can still achieve things as you're trying to figure it out. When you embrace imperfection, you have a bias towards action. And leaders and organizations need that bias for action to sense and respond to the conditions around them.

People who know they're imperfect, but don't simply accept that imperfection, as is, work hard. They're usually trying to bring their very best, and they're not letting perfection get in the way of better.

As a leader, you need the tools to help your people be great at imperfection and get over failures. One way to start building this toolkit is to look at your own failures. Where have you failed recently? Can you spot the imperfections in your life?

Where to put the tightrope

Maybe, right now, you're not even looking at your imperfection. Maybe you're not even looking *for* it. Because we're conditioned not to see it. You should see your imperfection, and then ask yourself, "What am I going to do about this?" Don't pretend not to see your imperfections, because then you'll never get better. But don't drown in them either.

As I mentioned, the best way to move forward is to make your potential failures small enough that you could still recover easily. If you want to walk a tightrope, set it just slightly above the ground. Then you can afford to fail. If you take your first tightrope walk off a fifty-story building, the cost of failure's just gone up.

Seeing your imperfections can also help reduce the cost of failure, because you're being transparent. If you take that practice walk on the tightrope and you're really wobbly, you should be able to see and feel that and decide what to do next. Maybe don't go for the high-rise building walk until you've worked out the wobbling. But if you're masking, and telling yourself you're doing it perfectly, you won't see the risk. And this goes for your people, too. Even your most trusted employees have flaws and bad days. If you allow yourself to see them as imperfect people, you'll be ready to spot those times when they need help.

The beauty of embracing your imperfections is, you keep going. That employee who is having a bad day—you don't send them home. You help them adapt and keep going. That way, they learn what it is to climb and get back up. The struggles you've gone through taught you how to get back up where you need to go. But don't be afraid to grab the next step, because you might fall down. Be hungry enough to keep doing those stairs.

Stay in experiment mode; always in discovery, always innovating. Because that's when serendipity can occur.

Walking—and falling—the 5K

Ten years ago I decided I was going to train to walk a five-kilometre race. Life was going well. I was exercising a lot, losing weight, and getting in shape. I had a trainer and I thought it would be a good time to do something. To give back.

I decided to do one of Canada's biggest cancer fundraisers—the Terry Fox run. I had spent a lot of time in the hospital as a kid, and I remember being in there and watching him on TV when the now-legendary Canadian was making his cross-Canada journey with a prosthetic leg to raise money for cancer. I remember thinking if I ever got out of the hospital and was healthy enough, I wanted to do something like he did to make things better for other people.

When I started preparing to walk for the 5K, I couldn't get past one kilometre. It wasn't the strain. I was just really scared to fall. Anytime I came close to losing my balance, I would quit for the day. Even after I started getting stronger and stronger, I couldn't break that threshold.

You see, before that I had never fallen while standing. Never during my whole adult life. So I was afraid to fall. You always fear the unknown.

Then finally, one day during training, I fell. It stung a bit. I was embarrassed. I thought for a moment, 'Maybe I can't do this.' But my trainer helped me back up and I got to do another kilometre and a half. I failed *and* I learned I could fall and get back up. In other words, I never really learned to walk 5K without falling.

Rather, I learned to get up from falling to make it to 5K.

It's not that the fear of falling was bad. It was a rational fear. People get injured from falling. But it was holding me back. It was telling me;

You're never going to recover from this. People are going to doubt you. Why should you even try this walk at all? You don't really need to, you've already overcome so much. Why don't you just live off yesterday's success?

But instead, I wanted to do more.

I can't say that I really embraced falling. It was one imperfection I could have done without. But it got easier. I got great at being imperfect, because if I had let perfection stand in my way, ironically, I would have never known what it was to do a 5K.

So my training had two components. One was getting in physical shape. The other was mentally preparing for failures along the way: falling and getting back up. And if I hadn't learned what it was to struggle up those stairs in my childhood home, I wouldn't have built that muscle memory. I wouldn't have learned that struggle is something you don't avoid, you should run at.

Learn to fall often

I didn't just learn to fall. I learned to fall often. Working with my trainer made it safe to fall, because I knew he would catch me and help me back up. We set the conditions where I could afford to fail, and then I could push myself further.

Stuff's going to happen. If you don't embrace failure early, you're delaying the process of learning yourself and how to recover.

Avoiding that failure is really cheating yourself. It would have been tempting to try to keep my perfect mask, train in secret and just present myself as a 5k athlete. But it helped make everyone else more comfortable when I fell. Because when somebody in a wheelchair falls, it seems so dramatic. People are like, 'Oh my God! He's dead!'

But I wasn't dead, of course. The route to the finish line was filled with countless falls and, with the help of my loyal and devoted team, I got back up and kept walking each and every time, raising $30,000 for cancer research in the process. Not bad. A decade leater I trained again after putting on weight. I trained, got in shape and raised over $85,000 to be Canada;s top fundraiser that year.

But, just like with that walk on the beach, once they saw I could get back up, it built comfort for everybody else that failure could happen.

And that's what inspired that older gentleman to walk to the water. Not because I made it to the water, but because he saw me fall and get back up.

"I just knew that if you could fall down and look like an idiot and get back up, I could too," he said.

I guess I shouldn't have been surprised, and neither should you.

After all, everybody knows the best stories are all about the comeback.

Exercises

1. Take an inventory of your imperfections within yourself and your life, including work. Start in the past, when you were younger. What imperfections do you have that you learned to live with, or you learned to move on from?
    a. What failures have you made recently? Can you spot the imperfections in your life today? Not the ones in the past, but the ones you're still facing?
    b. Look back at times that you've traveled through a failure or pain. Ask yourself, "Well, what did I do to get through that? What did I call upon within myself?" Because chances are, whatever you called upon within yourself was not unique to that situation. All those resources have come forward, and if you've gotten through it once, you can always get through it.

2. Name all the times where you had to ask for help to learn something, and then you got really good at it. How often did you offer to help someone afterwards?

3. Take stock of your inventory of imperfections and your resources. Consider, what imperfections and opportunities are available in your life and at work right now? Where in your personal life is there an experiment that you can do that won't cost them anything, to help them get used to this new challenge? How can you practice being resilient and laughing at yourself?

a. For example, say one of your children is getting married and you want to dance at their wedding. Well, that's an opportunity to learn to dance.

You could avoid it, because it might be embarrassing, or you could start small, by dancing at small gatherings where people won't all be watching and judging you.

Maybe dance in front of your daughter, because she'll laugh and think it's really cute. Don't do it on Dancing with the Stars.

b. Once you've tried this out in your personal life, take it to work. What imperfection could you work on where your employees can see it? Because that's how you model embracing imperfection.

Conclusion

I pretty much always suffer from imposter syndrome. It's uncomfortable, sure, but the nervousness that comes with it keeps me going. It excites me, and reminds me that I'm human, which is important in my day-to-day at Microsoft.

We put the human being at the center of everything we build—which requires hiring good-hearted people, of whom Microsoft employs many, and I swear I'm not just saying that because I work there. If you do the same thing, really at any company, your prospects for success rise exponentially.

But serving people—your customers—comes at the very end. Long before you arrive at that stage, you need buy-in from your team to build.

Microsoft is deep in the development of new technology that will utilize device cameras as an alternative input, allowing users to move their screen pointer with their eyes. Thanks to this innovation, people won't need a mouse or even a trackpad anymore.

If you've read this far, first off, thank you, but also you can probably guess that I played a role in its production. Even with an important, progressive and serviceable piece of tech like that one, I've needed to motivate teams to push it forward.

Who did I hit first? This time, the buy-in was bottom-up. I proselytized about the product's potential for global impact—not just on disabled people, but the entire population—to the designers and engineers. (I've often told people who asked me why accessibility matters, "We're all going to be disabled someday, some of us just beat you to it." *Wink*) This was a risk because they essentially had to work on the project in secret, on the side, without funding. My superiors didn't know what we were up to.

Some might've seen that as self-confidence coming through. I'd probably call it "stubbornness." But I say, *Don't let the org chart dictate how you should work.* Instead, work across it to do what's needed.

My Design team and I had all met people with disabilities who could benefit. But I also said to my workers, "How much would *you* love to be in front of your computer and not have to worry about reaching for a mouse?"

The team rallied out of passion. I'd giftwrapped them agency, and created a movement.

However, I was only in a position to even attempt such a stunt because I'd already cultivated such close relationships with my team members.

We built an awesome prototype in what seemed like record time—three months—and got it funded immediately. Talk about agility!

Then, a CVP came up to me and said, "I can't get my people to work together like that." I told him, "That's because you approach team leaders first. I went to the people who are actually doing the work."

As I continue on at Microsoft, I've learned that nothing is impossible. The emergence of artificial intelligence—which I predict will be the greatest enabler for people with disabilities we've ever seen—has only solidified that belief. What is becoming commonplace in the AI arena has had a tremendously positive impact on my quality of life. Take generative AI's ability to craft emails. That alone saves me, and countless other disabled people, hours upon hours of time each week. And the same tech helps me write reports and many other kinds of documents, as well.

How many more people like me are going to excel in a work environment and move up the corporate ladder because technology like gen AI has evened the playing field?

I'm already seeing more representation in corporations. Before I was at Microsoft, I'd attend leadership conferences and had no problem going to the bathroom right when I needed to. I was the unicorn in the conference hall, and there was never a line outside the accessible bathroom. Now, after a panel that bathroom's line is the longest of them all, because we take so much time in there! In all seriousness, though, the world is going to get even better with representation, partly because of technological developments, and that's awesome.

My message to disabled people with high aspirations: My disability has helped me become a better leader, so embrace yours. Get comfortable with it and find out how it *strengthens* you. For me, as a person with cerebral palsy, I've always had to depend on other people. Well, that gave me loads of time to learn how to motivate my helpers and get the best out of them. It was a necessity for me, and it's easy for me to forget sometimes that I even have a disability when I'm doing my thing at work. I often feel like my team of 12 is as productive as a team that's sized 12-squared, honestly.

Throughout my time at Microsoft, I think I've done a good job helping to amplify the importance of human factors and making that a priority in Design. After integrating Accessibility into that team, the creativity flowed inside out, which is far more impactful and inspiring.

We now build an emotional human connection with the technology. That's why I renamed the department "Human-Centered Design." That connectivity builds brand loyalty. People—disabled or not—won't just use our products, they'll choose our products.

But as I've outlined, there have been plenty of moments where I've tumbled, as have my teams. However, since that day I walked across the beach all the way to the water, I've thought about the power of imperfection. Once you embrace the idea that you're flawed and that you will make mistakes, you can unlock your true potential. Imperfection can be an enormous asset if you allow it to be. It widens your horizon—and out there are all your opportunities for growth.

When I turned around in the water and saw the world from a new perspective, I learned the benefit of deliberately looking at things from different angles. I learned that the way I see things isn't an absolute truth. It's just one perspective.

I also learned how easy it is to lose track of how far you've come. As I was walking into the water, I kept wanting to go further, worried that I hadn't gone far enough. But when I turned around and saw how far I'd come, I realized I'd done what I set out to do. It was mind-blowing and empowering. Inspiring.

That walk taught me the value of taking the time to reflect, to see how far I've come. From where I'd been sitting on the beach, the water looked far away. I really didn't know if I could make it. But from the water, it seemed like a trivial distance back to my beach chair. If I hadn't taken that time to turn around and reflect, I wouldn't have had a chance to appreciate the beauty of where I've been and where I am now.

And I urge you as leaders to do the same. Instead of always pushing forward and trying to walk perfectly, take time to look back and consider how far you've come and what you overcame to get here. Notice how different your original starting point looks, now that you're in a new spot. How did you feel then? How do you feel now?

Too often, leaders just look ahead without appreciating how far they've come. Pausing to reflect gives you a chance to understand your trajectory up to now, and to decide what your trajectory ought to be going forward. It also gives you a chance to share and celebrate your success with those who helped along the way.

When I crossed the finish line for my 5K walk and when I turned around in the water at the beach, I had loved ones with me to share those moments. Success can feel empty if you don't have a chance to share it with those special people around you. I hope this book allowed me to share the journey through failures and successes in a way that feels meaningful to you.

Who have you been sharing your experience with as you work through this book? Ideally, this should include your employees, peers, and maybe even people you report to. All of these people are with you on your journey now; they've taken the first steps with you and in some ways, it's their journey too.

When you start a new challenge, it can be tempting to keep it to yourself. But if you wait to share until you can walk perfectly, you'll never get to go on this journey at all. Waiting for perfection might mean you'll miss out on seeing what you can do, even if you follow an imperfect path to get there. Let the imperfection be part of your journey. Let it be something you work on improving as you go.

Starting a new journey will bring you new experiences, no matter what happens. The only thing holding you back is waiting for the "perfect" conditions. And nobody has time to wait. So be in discovery mode, let your team lead and allow yourself and your organization to experience imperfection. And let others see you do it, because experiencing imperfection allows you to find new ideas and opportunities faster.

This journey to leading with imperfect feet isn't something you should do alone. Rather, you should be doing it *with* people, side-by-side. Leading isn't always being out front, it's being able to move everybody together, and let them help you lead, because no leader can do it alone.

If you don't have a person helping you hold up and walk, and move forward, you won't be able to persevere.

All of my successes were achieved by working with people, not going around them or using them. As a team, we shared a goal and advanced together, step by step.

It worked on the beach, it worked on the Terry Fox run and it works every day in organizations where I demonstrate to teams the power of moving hand-in-hand and encouraging each other to learn from failings and get back up after a fall.

Believe me, if you build the right culture and environment, the people who you serve will be the first to pick you up. And if we can stay in that space, where we can pick each other up and help one another through our imperfections—falling, dusting ourselves off and getting back up—there's no telling how far we can go as a group.

That's why you need to build relationships, bonding not only with the people who report to you, but also with the people all around you. Create relationships where you're not looking to outdo each other, but to bring everybody and each other forward. That's how we walk imperfectly; because if we're all imperfect together, each person's imperfections matter less.

Exercise

There's one last exercise I'd like you to do. It's to go back and review all of the exercises you did in this book. Time has passed since you started this journey, and it's time to see how far you've come. As you go back through the exercises, notice how differently you think now compared to the first time you worked through them. Notice how you've grown and changed. If you were to do the exercise again right now, what would change?

When you arrive back at this section after your review, take a moment to embrace your imperfect feet and how far they've already taken you.

Now ask yourself: **What's the next imperfect step?**

Dedication and acknowledgments

I'd like to acknowledge the following people for the huge role they've played in my life.

The love of my life, Kelly

As I mentioned in Chapter 4, Kelly and I met online, which was weird at the time. I was working for one of my first employers, who's also acknowledged here: Rob Whent. We were creating banners for an online dating website and they didn't have a test environment, so I had to do the testing in production, which meant I needed free credits to actually use the website. I was dating, but I hadn't found that special someone. So I figured, it's free anyways—I might as well try it. I made a profile and included the fact that I had cerebral palsy. I met Kelly, and we started instant messaging back and forth. Typing isn't that easy for me, so I asked her to move to phone calls.

Once we were on the phone, the third question out of her mouth was, "Can you have sex?" I talk about this more in Chapter 4, but essentially, I loved how up front and honest she was. It gave me permission to be honest too. We didn't get off to a great start, but we came through it and we've been happily married for 15 years.

What I really value most about my wife is that with her, I can take the mask off and be completely vulnerable. I can show her my true self, at times when the rest of the world has been beating on me. She's an extrovert and I'm an introvert, which is great. She can tell when I'm done being in front of everybody, and that's when she swoops in and deflects the attention and conversation. That gives me a chance to recharge and do what I need to do to be my best me.

Living with a disability means you're conditioned to not show weakness, to be strong. You have to push through challenges and overcome them. That's what makes you unique and special. It can feel like you can never be down. But with Kelly, I love that I don't have to mask. I can let her see the real me.

We've grown together over the years. We're both probably misfits in our own way. But we attract each other, and we get to radiate our unusual selves together.

We moved to a new city so I could grow my career and take on new experiences. Kelly sacrificed so much for me to do that, so I could build my presence with public speaking, like my TED talk. When I get caught up in something, I go all in. Kelly gives me the space and time to pursue those things. She's always been a supporting, caring partner, celebrating with me in my victories and shielding me when I had nothing left to give.

When most women imagine their Prince Charming, they don't usually think of someone with cerebral palsy. Prince Charming rides a horse, he sweeps you off your feet and you walk off into the sunset together. But Kelly's Prince Charming happens to be in a wheelchair. She didn't know what it'd be like to be with somebody with a disability. But she made the choice to be courageous and take a chance on me, to be romantically involved and intimate with me, and decide with me to not have kids, so we can enjoy and appreciate each other. She's chosen to be there and help me have the life that I have, even though I don't know how long I'm going to be on this earth. I can't thank her enough for sharing this journey with me. I wouldn't want to take a step on this very rocky road in life with anyone but her, because with her I'm whole. The wheelchair helps me get around, but she helped me get into life, to really experience what love can do, and to connect deeply with her.

When I'm no longer here on this earth, my one wish for Kelly is that she continues to live and move on. That she goes out and meets somebody who will give her what she gives me—and that's everything. That she gives that heart of hers to others, and that others give her their hearts. So she can live every last day being loved by someone as I love her, up until my last day.

With her, I don't have a disability. With her, I just get to enjoy. And we don't say it nearly enough to each other, but I want her to always be able to refer to this book and read this acknowledgement. Thank you, Kelly, for allowing me to live a life that put words on these pages.

These words would've never been formed if she didn't walk step by step with me as we moved cities, changed jobs, and as I pursued my passions and got all caught up in whatever I was trying to overcome. She gives me that space to do that, when not every partner would.

While we watched *the Defiant Ones* music documentary one night and Doctor Dre was being interviewed, Kelly said, "You can't marry somebody like Dre unless you appreciate what kind of artist he is. Because he'll go away in the studio and he'll be gone forever, because he's so into what he does. You don't marry somebody with that kind of curiosity, passion, interest, and not know that you're going to be sharing them with the rest of the world."

I'm thankful that Kelly shares me with the rest of the world, so I can have and do what I do. She sacrifices time with me so I can pursue these endeavours and be there for other people when I can't always be there for her. Without her, I couldn't be that person who could do both.

Team Dave: Al and Ida Dame

I can't thank my parents enough for the tough love, stubbornness and devotion that it took back in the 1970s for them to envision a world where somebody with cerebral palsy could have a fulfilling life. They embraced the discomfort of figuring out what that life could look like. And then they ruthlessly pursued their vision by challenging the school board processes and societal norms to ensure I could be accepted as a person with a disability.

I am still realizing all of the effort and sacrifice that they made for me, and I'm newly thankful every day. They made me who I am, and they're the reason I've made it to where I am today.

I only wish they could see me now. My parents have passed away.. I hope somehow they knew their sacrifice was worth it. They worked extra hours to afford the things we needed. Our family income was high enough that we didn't receive much government assistance, but they had to make financial sacrifices to cover the expenses that came with my disability. I think back to the many long hours my parents spent with me in the hospital when I was a child.

During my worst times, when I was in pain and I couldn't imagine life would be like after all the surgeries, they were there for me. They made sure I didn't dwell on what was, and reminded me to think about what life could be.

People always say to me, "You fought for your life," because I was expected to die, and I fought to live. But it was my parents who started the fight, making sure I could attain opportunities and a fulfilling life, even when no one knew what that would even look like.

There was no clear path for my parents to follow back when I was born, no social scripts for parenting a kid with cerebral palsy. Most new parents can emulate what their friends do for their kids, or what their own parents did for them. But in the 1970s and '80s where we grew up, there weren't many models for families with disabled children. So my mom and dad struck out on their own to figure out what someone like me could do. When I think about it, they were role models for the leadership approaches I talk about in this book. Take changes, try new things, let your learners teach you and be willing to fail to achieve your goals.

My parents wouldn't have had access to the kind of thinking I now encourage back then in their social circles. We were a blue collar family. My dad owned a heating and cooling company and was a hands-on leader. At his funeral, many people told me stories of his kindness. How he helped buy them groceries when they weren't working, how he fixed their roof, how he did free boiler repairs for a decade to help a church that couldn't afford to pay him. When he died, they sent us a cheque for $100. "This doesn't even come close to what your dad did," the note said.

I always wanted to be a leader like my dad. His employees respected him. And he created a culture where all these very different characters were brought together in a great group. I don't think he even really thought about it that way—it was just how he was.

My mom was probably just about ready to kill me at times, and with her stubbornness and her five-foot-two, firecracker personality, she could fight above her weight in battles with more endurance than I've ever seen since.

She didn't treat me special. She spanked me when I needed to be spanked. I remember when I was a kid on an errand to the bank with my mom, I was waiting in the car and thought it'd be funny if I locked the doors. It was pouring rain. When she asked me to open the door for her to get in, I just silently shook my head no, all mischievous. "David, you better open this goddamn door!" she said. "Nope." Then she said, "When I get in there I'm going to blister your backside." So of course, I knew I wasn't letting her in after that. And finally, she told me this: "If you let me in, I promise not to spank you." So, naively I unlocked the doors. She sat down in the car, undid my seatbelt, pulled me over her lap and spanked me then and there. I started to cry. "But you promised!" And she said, "Sometimes grown ups lie." That was a lesson learned. And I deserved it. She taught me that my disability would not hold me back to achieve anything, nor would my disability allow me to get away wih anything.

My dad was the calm one and mom was more the bad cop: ready to challenge and do what was needed.

David Bellamy

I met David Bellamy when I was four years old. He was my next door neighbour and four years older than my other friend Danny and me. But there weren't many kids around since it was a rural area, so we played and hung out. Growing up, my dad would pay him to wash the car and truck. And because he was older, he was my parents' go-to babysitter.

Dave is the older brother I didn't have. We fought, we hated each other at times and then were back to loving each other. We played video games together. Dave would bring me to swimming—gym on Saturdays; all those things that a big brother would do. But as I got older, in my teenage years, he helped me get into trouble. He showed me cool things like house parties—so I could really experience living, not just being cooped up at home all the time.

Dave was there to give me support and push me into tougher courses at school. He was there to help me find acceptance and friends. He made sure I went to high school dances. Where my parents fought for accessibility, Dave helped me be included. It was one thing to get me admitted to the school and get accommodations for me to take classes. But it was a completely different thing to actually engage with the other students, and Dave helped make that happen. He would go to high school dances with me and because he was cool, it helped. Even when he was in college, he'd come to those dances. Getting a college person to bring me to the high school dance made me feel pretty confident—it was life changing to have this older, cooler person to hang out with.

But Dave would also push me. When my parents would let me get away with things, he would step in and call bullshit. I needed that at times. He was there through all the special moments in my life, making sure I was doing okay. I can never thank David enough for being my advocate when it came to helping me be included in things. From when I was a kid all the way to when I went away for university. Especially through the teen years, even bringing me to bars once I was legal. Because bringing disabled people to bars probably wasn't the coolest thing, but he did it so I'd get to have that experience.

Even when his life got busy and he became a dad at 20, he still found time to be with me and help me in different parts of my life. When I started school, he would help get me ready during the day or stay with me while my parents went away for the weekend. When I went to high school, he made sure I would fit in. When I went to university, he helped me move and settle into my room. After we graduated, we reconnected, even though I went out of town for school. And we reconnected again during the years when my dad passed away. He was always someone I could lean on. We attended each other's weddings.

I spoke at a cerebral palsy gala recently, and he came, so I was able to thank him. We've been through so much together. He has a wife, a daughter and now he's a grandfather. And it's incredible to see that we're still close, still a part of each other's lives.

Bourilack Phommasenh—"B"

When I met B, he was an 18-year-old punk kid. He was doing personal care because his mom and cousin got him into it. He would get in trouble like most 18 year olds, but he had a job that came with a huge responsibility. We met through my friend, Jeff, who I knew from university. Jeff was in a wheelchair, and recommended B, who was one of his helpers.

B is the kind of guy who's always looking for a shortcut or loophole, some way to scam the system. When we started working together, I didn't know how long I would keep him, because generally speaking, by that point in my life, I had found that men weren't as reliable as women. I was skeptical if he'd be reliable enough for the level of support I needed. But he was reliable, and we've been working together ever since.

We spent many years forming a strong relationship. Just as Dave was a big brother to me, B became like the little brother I never had. He started handling my care right when my career was starting to take off. I was getting recognized professionally for the organizational change work I do, and starting to travel. That was relatively new in the corporate world, where there wasn't a huge percentage of disabled professionals who needed to travel. My company didn't even know they had to hire a helper to go with me (I had to request travel reimbursement for him too). But B was right there to help me be the best me. Even though he had a job that people really look down upon, he was dedicated, because he loves taking care of people.

B showed me that the world is a good place and it has some remarkable people who don't want anything for themselves, but who truly make the people around them better. You can drop him in front of any group and see his smile and way of being just radiate through a room. He's real and honest; someone everybody strives to be. Even though what he does isn't glamorous, he's made countless people's lives better. Mine, specifically. I wouldn't be able to speak in front of all the people that I have, corporately or professionally, if he wasn't so perfectly behind me.

I probably wouldn't be where I am if it wasn't for the care B gave me, the devotion, and the fun times we had. When you travel with somebody, you really get to know them. We would spend many hours on the plane or in a hotel playing practical jokes on each other, bonding and connecting. The jokes we played on each other could be downright ruthless, but that's how you know you're close, when you can be like that to each other.

Now, what made B such an integral member of my team is that he knows when to give me my space and when to be a shadow, and when to jump in when I need the help. He is so tuned in to my needs, he knows how to be invisibly present—which is the mark of a true leader.

While providing support, he has never tried to take the limelight onto himself and what he was doing for me. He always gave me care in such a way that people would forget I had a disability.

Even I would forget I have a disability. He helped from a distance, letting me strive unhampered, to do what I could do. He is somebody that I try to model myself on as a leader, and coach other leaders to do the same. Because a leader is somebody who can help employees bring their best selves. And that's what he does every day. He is one of the best leaders I know because he's willing to do whatever it takes to help people bring their best selves to the world. That's a true leader.

Today, B has three kids, and he's finally married. I For everything he's been, for everything he'll be, and for everyone who will ever get to experience him. He makes the world a better place. He is what the world should strive to be.

Here's a story of a joke I played on B. He had never flown in a plane until he flew with me. We were heading to Florida for a conference where I was going to speak. B was all excited. We got to our seats on the plane, seat belts on. B turns to me and says, "This is going to be so great. What do I need to do?" Meaning, what care did I need during the flight. So I said, "Well, when the stewardess comes by, if I was you, I'd ask her where the double parachute is." He said, "What?" I said, "Well, if the plane crashes, I can't pull my own parachute. So, you've got to strap me onto your back and we need a double parachute."

B was suspicious. "Really?" He asked. I said, "Why would I make that up?" So, when the flight attendant came by, B said, "Excuse me, ma'am. Where do you keep the double parachute?" Of course, she said, "What?" So B explained, "Well, Dave here is in a wheelchair, so if the plane goes down, I've got to strap him on my back and we've got to jump off the plane."

The poor flight attendant was looking at him with a blank stare on her face, and her mouth open. She looked over at me. I shrugged and said, "I don't know where he gets this stuff from." Relieved, she moved on. B said, "You son of a bitch." So I said, "Sorry, I couldn't resist." He really is like the younger brother I never had.

Andrew Wertkin

My old boss Andrew helped me understand what was possible for me as a change agent. He really got it through my head that I was bigger than my job, that I could have a bigger impact in the world than other people because of my story and my approach to life. He could see a professional vision for me that I couldn't see myself. What Andrew did for me was like a continuation of my parents' work. They knew at an early age that I could live a bigger life outside an institution. Andrew saw how I could market and sell my vision, grow my career and travel the world.

To this day he stays connected with me, because he's stubborn like that. He's got that New York in-your-face bravado. It's fueled his success, because he can have a vision that others don't and just push forward with it anyway. That's what made him a successful entrepreneur and C-level of many organizations, and he taught that to me, too. He'd always say, "F*ck what other people think. Go to the vision and do it." That's made him a successful entrepreneur and CEO leading organizations.

We met when he was working for a company that was trying to recruit me and I give him credit for having the patience to try to poach me for a year. He made me an offer, I turned it down, and he waited months. I remember our first interview when we finally sat down together. He asked, "How have you been so successful?" And I said, "It's because I've made a lot of mistakes." That struck him as a really profound thing to say. He said later it made him realize what kind of person I was. When I finally joined the company, another company bought us and moved him from Toronto to Boston. So for the first few years that I knew him, he wasn't my direct boss. But he was an incredible mentor.

I would always think small, and he would push me to think big. *Think big, think beyond that*. He would continually push me beyond the walls I was imposing on myself.

Andrew felt uncomfortable that he recruited me, then moved to Boston. But every time I would fly to Boston for work, we would connect and continue to get to know each other. And he really helped me move up in the company by telling people what kind of change agent I am, sharing my professional successes. He often talked about how I was able to connect and relate to people around me and how I could get people to want to do things instead of having to be told to do things through the chain of command. He was behind the scenes as my biggest advocate.

Then he decided to leave the company. When he told me, Andrew said, "I want to tell you that I'll be leaving the company in a couple months, but you're one of the best two hiring decisions I ever made." And I thought, *wow, thank you*. That's great. No harm, no foul. We did a lot of great things.

Three months later, he messaged me: "Hey, I'm in Toronto." So I asked, "Are you trying to escape the draft? Why did you move to Canada?" Turns out, he had started at a company and he wanted me to help change the culture.

I was living in Waterloo, about an hour and twenty minutes' drive from Toronto. Andrew wanted to know if I'd be willing to move to Toronto. I was training for my 5K walk at the time, so I said no thanks. Andrew came out on the day of the walk and walked with me, and held me up as I was doing this thing that other people told me I'd never be able to do. Andrew was never one of those people. He always told me I have an ability to change the world around me. As he said, "Other people talk about change, but when you struggle and do things, it's more than any poser saying, 'I struggle.' Because everybody can see your struggle and what you overcame. And you have an ability to influence the world around you. You can change the world and become really successful, but in Waterloo you will struggle to radiate." He also knew my wife had been given notice that she was being laid off. So he asked, "What's keeping you here?" He offered me a change, to come work for him, and gave me what I needed to establish myself in Toronto.

Every so often we reconnect now. He gets to see where I am and I get to thank him for giving me the opportunity, and pushing me to be bigger than I could ever see myself. To achieve things professionally that I never thought I could. He did that.

I don't know where I'll end up, but wherever that is, I'll end up there because Andrew believed in me, and continues to connect with me and be part of my journey.

Panos Panay

Panos Panay -my forever gratitude goes to you. Your confidence in me gave me the courage to imagine bigger than I ever thought possible and to relentlessly drive product making that truly connects with customers. You showed me that compassion, empathy, and human connection are not soft skills but the foundation of meaningful innovation. Beyond leadership, I am grateful for our friendship. The trust, encouragement, and laughter we've shared remind me that great products are born not only from vision, but from genuine human bonds. Your belief in me has shaped both my work and my life.

Kris Hunter

To Kris Hunter. You are the one who interviewed me and made the decision to hire me into Microsoft your decision on me at a moment that would ultimately change the direction of my life. Your belief went beyond a role or a resume. You saw something, trusted it, and acted on it.

Kris, you are one of those rare people who truly change the course of someone's life. I am deeply grateful for the opportunity you gave me and for the impact that decision continues to have on my work and my journey.

Marcus Ash

To Marcus Ash. Thank you for elevating me to your leadership team and giving me the space to show what agile product making can be. You saw me as more than the accessibility guy. You knew I was a product maker. I am grateful for the many engagements where we learned from each other.

Your belief in me created a lifelong bond. Cerebral palsy took my ability to walk, but your trust gave me the ability to fly and soar. It shaped how I lead with authenticity and how I help others rise, just as you helped me.

Aaron Sampson

Aaron was my first corporate hire in Toronto when I was working for Andrew. It was the first time I got to see more in somebody than they saw themselves. When he came to interview for a job; he didn't really have the skills or experience, but I could see in him what he couldn't see in himself. I took a chance in hiring him. He had a strong work ethic, and I used that to help him grow into a phenomenal manager and professional—better than he thought he could be. He's achieved great things, in a very humble way.

I got to do to for Aaron, what others did for me. He was there in Toronto when I was feeling pretty lonely. We got to really grow together. I had started out being his mentor, and the one that leads him. But over a four year period, we got to the point where I learn from him probably as much as he learned from me. He's another little brother I wish I had; I see so much how he impacts people, and makes everything around him better, which has made my life better in Toronto. He coached me through my whole Ted Talk to make sure I had the story just right.

Usually when you invest in somebody, it's one way. You don't usually get two-way value. But Aaron is somebody who I invested in and I got back my investment three-fold. He invested in me too, helping me to become an even better, more refined version of myself. Even though I'm one of the best in what I do professionally, he helped grow me into a better, more known and popular professional. He got me to really feel comfortable with the impact I had and the role I was going to have.

Jordana Feldman

To Jordana Feldman. Thank you for helping me shape the title of this book and for framing the Spark presentation that became the foundation for my story. You helped me see the narrative more clearly and gave structure to ideas that had been living in my head for years. Your insight and guidance played a meaningful role in bringing this book to life.

My Writing Team:

To Michael Stahl, Joseph Quaderer, and Greg Shaw. Thank you for working so closely with me to turn my ideas into a book. Your commitment and belief made this a reality. I am grateful for your guidance, your patience, and the craft you brought to every step of the journey.

Thank you to Scrum.org, Easter Seals Camp Woodeden, the John McGivney Children's Centre and all my close friends that made, and friends that I consider my tribe who round me out to be who I am.

I started working with **Scrum.org** in Boston around the time that Agile was becoming more mainstream. Together we co-created the Nexus, a scaling scrum framework. They're a big global community with trainers all over the world. They gave me a vehicle to be able to share my work and inspire people globally. They're like a family to me, in that I hold my relationship to everybody in that community sacred. Because they want to do things for the right reasons; they want to help organizations harness change for a competitive advantage. They're so willing to find their success through others; they represent me and everything I want to be. They allowed me to have a global voice.

If it wasn't for my experiences at **Camp Woodeden,** I would've never ventured outside of Windsor. The camp had able-bodied counselors who would help me, and it made me realize that I wanted to move out of Windsor to go to school. They taught me that there were good people out there who would work with me, and gave me the confidence to move away from home. Some of my counselors are lifelong friends to this day. One of them attended my wedding.

Finally, my close friends are my tribe, because you get older and your circle gets tighter. That's a true circle. I want to thank them without mentioning their names. They know who they are.

About the Author

David Dame is a globally recognized product leader, accessibility champion, and change agent with more than 30 years of experience guiding organizations through complexity and transformation. Born with cerebral palsy, David learned early that progress is built through experimentation, vulnerability, and the grit to keep moving forward one imperfect step at a time.

He has led large-scale enterprise agile transformations, trained hundreds of leaders worldwide, and brought inclusive innovation to life through products like Microsoft's Adaptive Accessories, the Surface Adaptive Kit, and other multimodal technologies that help the world work more equitably. David is also a TEDx speaker whose talk, *Sprinting with Cerebral Palsy*, has inspired audiences around the world.

Today he serves as **Senior Director, Human Centered Design at Microsoft**, where he helps shape the next generation of inclusive products by blending user imagination, business strategy, and his lived experience navigating a world not designed for him. His work demonstrates that embracing imperfection is not a compromise—it is a competitive advantage.

Other titles from 8080 Books

No Prize for Pessimism by Sam Schillace

A Platform Mindset by Marcus Fontoura

WorkLab by Collette Stallbaumer

Human Agency in a Digital World by Marcus Fontoura

Hyderabad Days by Ravi Vedula

Showstopper! by G. Pascal Zachary

Check for new titles on our website:

https://unlocked.microsoft.com/8080-books/